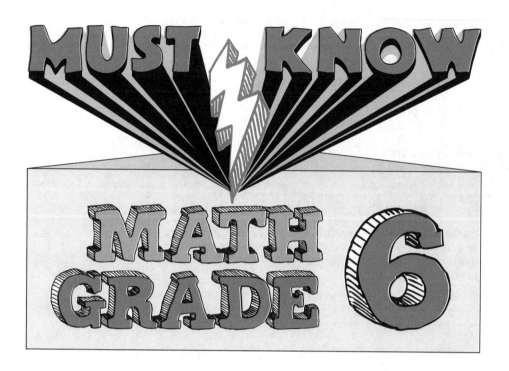

Nicholas Falletta

New York Chicago San Francisco Athens London Madrid
Mexico City Milan New Delhi Singapore Sydney Toronto

Copyright © 2020 by McGraw Hill. All rights reserved. Printed in the United States of America. Except as permitted under the United States Copyright Act of 1976, no part of this publication may be reproduced or distributed in any form or by any means, or stored in a database or retrieval system, without the prior written permission of the publisher.

1 2 3 4 5 6 7 8 9 LCR 25 24 23 22 21 20

ISBN   978-1-260-46408-5
MHID   1-260-46408-3

e-ISBN   978-1-260-46409-2
e-MHID   1-260-46409-1

Interior design by Steve Straus of Think Book Works.
Cover and letter art by Kate Rutter.

McGraw-Hill Education products are available at special quantity discounts to use as premiums and sales promotions or for use in corporate training programs. To contact a representative, please visit the Contact Us pages at www.mhprofessional.com.

Contents

Introduction	vii
The Flashcard App	ix
Author's Note	xi

1 The Number System — 1

The Base-10 Number System	2
Different Types of Numbers	4
Using a Number Line	10
Absolute Value	13
Performing Operations with Whole Numbers	14
Properties of Numbers	19

2 Decimals — 27

Understanding Decimals	28
Reading and Writing Decimals	30
Comparing and Ordering Decimals	33
Rounding Decimals	34
Adding and Subtracting Decimals	35
Multiplying Decimals	42
Dividing Decimals	46

3 Fractions — 53

Understanding Fractions	54
Multiples and Factors	56
Equivalent Fractions	61

Adding and Subtracting Fractions … 63
Adding and Subtracting Mixed Numbers … 65
Multiplying and Dividing Fractions and Mixed Numbers … 67

4 Integers … 77
Understanding Integers … 78
Adding Integers … 80
Subtracting Integers … 83
Multiplying Integers … 84
Dividing Integers … 86

5 Ratio and Proportion … 91
Understanding Ratios … 92
Finding Equivalent Ratios … 93
Comparing Ratios … 94
Understanding Rates … 96
Solving Proportions … 97

6 Percent … 103
Solving Percent Problems Using Models … 104
Relating Fractions, Decimals, and Percents … 108
Finding the Percent of a Number … 110
Finding What Percent One Number Is of Another Number … 112
Finding a Number When the Percent Is Known … 113
Finding Percent of Increase or Decrease … 114
Finding Simple Interest … 116

7 Equations and Inequalities … 121
Defining *Equation* and *Function* … 122
Reading and Writing Equations … 123
Solving Equations … 124
Graphing Equations … 128

Monomials and Polynomials — 130
Adding Polynomials — 132
Multiplying Monomials — 132
Multiplying a Polynomial by a Monomial — 134
Solving Inequalities — 135

8 Measurement and Geometry — 143

Customary Units of Length — 144
Customary Units of Weight — 146
Customary Units of Capacity — 148
Understanding Units of Time — 150
Metric Units of Length — 153
Metric Units of Mass — 155
Metric Units of Capacity — 157
Finding the Perimeter of a Polygon — 159
Finding the Circumference of a Circle — 162

9 Plane Geometry — 171

Identifying Points, Lines, Rays, and Line Segments — 172
Identifying Angles — 174
Finding Angle Measures — 177
Classifying Triangles — 179
Finding Unknown Angle Measures in Triangles — 182
Classifying Quadrilaterals — 183
Finding Unknown Angle Measures in Quadrilaterals — 184
Identifying Other Common Polygons — 186
Finding Unknown Angle Measures in Other Polygons — 187
Identifying Congruent and Similar Figures — 189
Identifying Lines of Symmetry — 192

10 Geometry: Area and Volume — 205

- Area of Plane Figures — 206
- Finding the Area of Quadrilaterals — 207
- Finding the Area of Triangles — 210
- Finding the Area of Circles — 212
- Surface Area of Solid Figures — 214
- Finding the Surface Area of Rectangular Prisms — 215
- Finding the Surface Area of Cylinders — 218
- Finding the Surface Area of Cones — 220
- Volume of Solid Figures — 222
- Finding the Volume of Rectangular Prisms — 222
- Finding the Volume of Cylinders — 224
- Finding the Volume of Cones — 226
- Finding the Volume of Spheres — 227

11 Probability — 237

- Defining Probability — 238
- Simple Probabilities — 238
- Sample Spaces and Tree Diagrams — 242
- Combinations and Permutations — 244
- Probability of Independent Events — 247
- Experimental Probability — 252

12 Data and Statistics — 259

- Defining Statistics — 260
- Measures of Central Tendency — 260
- Pictographs — 263
- Bar Graphs — 265
- Circle Graphs — 268
- Line Graphs — 271
- Stem-and-Leaf Plots — 274
- Identifying Misleading Statistics — 276

Answer Key — 285

Introduction

Welcome to your new math book! Let us try to explain why we believe you've made the right choice. You've probably had your fill of books asking you to memorize lots of terms (such as in school). This book isn't going to do that—although you're welcome to memorize anything you take an interest in. You may also have found that a lot of books make a lot of promises about all the things you'll be able to accomplish by the time you reach the end of a given chapter. In the process, those books can make you feel as though you missed out on the building blocks that you actually need to master those goals.

With *Must Know Math Grade 6*, we've taken a different approach. When you start a new chapter, right off the bat you will immediately see one or more **must know** ideas. These are the essential concepts behind what you are going to study, and they will form the foundation of what you will learn throughout the chapter. With these **must know** ideas, you will have what you need to hold it together as you study, and they will be your guide as you make your way through each chapter.

To build on this foundation, you will find easy-to-follow discussions of the topic at hand, accompanied by comprehensive examples that show you how to apply what you're learning to solving typical sixth-grade math questions. Each chapter ends with review questions—250 throughout the book—designed to instill confidence as you practice your new skills.

This book has other features that will help you on this math journey of yours. It has a number of sidebars that will either provide helpful information or just serve as a quick break from your studies. The BTW sidebars ("by the way") point out important information, as well as tell you what to be careful about math-wise. Every once in a while, an IRL sidebar ("in real life") will tell you what you're studying has to do with the real world; other IRLs may just be interesting factoids.

In addition, this book is accompanied by a flashcard app that will give you the ability to test yourself at any time. The app includes 100-plus "flashcards" with a review question on one "side" and the answer on the other. You can either work through the flashcards by themselves or use them alongside the book. To find out where to get the app and how to use it, go to the next section, "The Flashcard App."

We also want to introduce you to your guide throughout this book. Nicholas Falletta is a veteran education writer and the author of McGraw Hill's *SSAT/ISEE*. We're glad to have the chance to work with him again. Nick has a clear idea what you should get out of a math class in 6th grade and has developed strategies to help you get there. He's also seen the kinds of pitfalls that students can fall into and is an experienced hand at solving those difficulties. In this book, Nick applies that experience both to showing you the most effective way to learn a given concept and how to extricate yourself from any trouble you may have gotten into. He will be a trustworthy guide as you expand your math knowledge and develop new skills.

Before we leave you to Nick's sure-footed guidance, let us give you one piece of advice. While we know that saying something "is the *worst*" is a cliché, if anything *is* the worst in the math you'll cover in this grade, it could be working with equations. Let the author introduce you to equations and show you how to work confidently with them. Take our word for it, learning how to handle equations will leave you in good stead for the rest of your math career—and in the real world, too.

Good luck with your studies!

The Editors at McGraw Hill

The Flashcard App

This book features a bonus flashcard app. It will help you test yourself on what you've learned as you make your way through the book (or in and out). It includes 100-plus "flashcards," both "front" and "back." It gives you two options as to how to use it. You can jump right into the app and start from any point that you want. Or you can take advantage of the handy QR codes at the end of each chapter in the book; they will take you directly to the flashcards related to what you're studying at the moment.

To take advantage of this bonus feature, follow these easy steps:

Search for **McGraw Hill Must Know** App from either Google Play or the App Store.

↓

Download the app to your smartphone or tablet.

↓

Once you've got the app, you can use it in either of two ways.

Just open the app and you're ready to go.	Use your phone's QR Code reader to scan any of the book's QR codes.
You can start at the beginning, or select any of the chapters listed.	You'll be taken directly to the flashcards that match your chapter of choice.

Get ready to test your math knowledge!

Author's Note

Must *Know Math Grade 6* builds on the skills and concepts that you learned in your elementary school math classes. Examples of clear-cut problems with complete solutions are provided in all the basic area of mathematics you'll need for success in this school year.

In addition to providing a thorough review of all the fundamentals of arithmetic that you studied through grade 5, you'll be introduced to many new skills and concepts about topics such as integers, ratio and proportion, percents, geometry, statistics, and algebra.

Mathematics, in general, and this book in particular are not meant for you to read only. This book contains plenty of example problems with complete explanations of their solutions that will guide you through the steps involved. Always try to do the problems on your own before you read the explanations. This process will give you a reasonable benchmark of how much you already understand, and it will clarify what you are having problems understanding.

After trying an example problem, check the solution. Even if you got the correct answer, look to see that you solved the problem in the best and most efficient way. If you have stumbled getting the right answer, remember that's why you bought this book. Read through the explanation and then close the book and try to solve the problem again.

At the end of each chapter, you'll find plenty of new problems in the Exercises section. These problems will make sure you consolidate and apply your understanding. Once you work through a set of problems, check the Answer Key at the back of the book. You'll be pleasantly surprised because it gives you more than just the right answers. For each problem, you'll find a detailed solution based on the model explanations offered in the chapter examples. The book also offers more than 100 electronic flashcards tied to all the lessons that you can access online.

Each chapter of the book begins with statements about what you **must know** when you have completed the chapter. After working through a chapter, if you don't feel that you have a good grasp of the concepts described at the beginning, make a point of going back through the chapter until you do. The more you are willing to do this, the more skilled and confident you'll become!

The Number System

MUST KNOW

- The base-10 number system uses ten digits: 0, 1, 2, 3, 4, 5, 6, 7, 8, 9.

- Integers include all positive numbers, zero, and all negative numbers.

- Rational numbers can be represented as fractions. Irrational numbers cannot be represented as fractions.

- The absolute value of a number is its distance from 0 on the number line.

- The properties of numbers explain how real numbers work together.

ur number system uses ten digits to form all numbers—0, 1, 2, 3, 4, 5, 6, 7, 8, 9. Called the base-10 number system, it dates back more than 4,000 years and is used throughout the world. Most researchers believe that the base-10 number system is so commonplace because humans naturally count on ten fingers. This makes the number system easy-to-use. You might even call it *handy*!

The Base-10 Number System

In the base-10 number system, the value of a digit is based on its position, or place, in the number. In the number 2,196,935, the 6 represents 6 thousands, or 6,000. The 3 represents 3 tens, or 30. What is the value of 9 in this number? Well, that depends on which 9 you are referring to. The digit 9 that comes before the comma represents 9 ten thousands, or 90,000. The digit 9 that comes after the comma represents 9 hundreds, or 900. The place value chart below makes it easy to identify the value of each digit in 2,196,935.

Millions	Hundred Thousands	Ten Thousands	Thousands	Hundreds	Tens	Ones
2	1	9	6	9	3	5

Notice that the value of a digit depends on its place or position in a number and is based on powers of ten. That is, each digit has a value that is 10 times the value of the digit to its right. Once the digit in a place is greater than 9, you must move left to the next highest place value. Numbers that appear to the left of the decimal point are greater than 1, whereas those to the right of the decimal point are fractions of a whole, or less than 1.

Place value charts are helpful in understanding the value of digits in decimal numbers. Consider, for example, the number 3,271.685. It's a lumpy kind of number, but you can get a good sense of what each digit means by properly recording it in a place value chart.

Thousands	Hundreds	Tens	Ones	Tenths	Hundredths	Thousandths
3	2	7	1 .	6	8	5

Notice the headings for each place to the right of the decimal point. Each has a unique name—*tenths*, *hundredths*, and *thousandths*—that ends with the suffix *–th*. This naming pattern continues in both directions. Let's look at an example.

EXAMPLE

▶ What is the value of 8 in the number 2,857.194?

▶ Identify the place name that corresponds with the digit 8. The digit 8 appears in the hundreds place.

▶ 8 in 2,857.194 is in the hundreds place and represents 800.

Now, let's try finding the value of a digit after the decimal point.

EXAMPLE

▶ What is the value of 3 in the number 58,209.693?

▶ Identify the place name that corresponds with the digit 3. The digit 3 appears in the thousandths place.

▶ 3 in 58,209.693 is in the thousandths place and represents $\frac{3}{1000}$, or 0.003.

Different Types of Numbers

When we first learn to count, we start out with the number 1 and count up: 1, 2, 3, 4, 5, 6, 7, 8, 9, 10. Not surprisingly, these numbers are called the **counting numbers** or **natural numbers.** Where does the number 0 fit into this system? The answer is that 0 and all the other positive numbers form a different set called **whole numbers**. The whole numbers will get you to infinity—if not beyond!—since they are unending. All you have to do is add one more unit to the last number and keep doing it—forever!

One way to represent whole numbers is to use a number line. A **number line** is a line with numbers marked in equal intervals. Number lines are helpful when you are comparing the value of numbers. For example, we can use a number line to compare two numbers such as 2 and 7 to determine which is greater than the other. Since 7 is to the right of 2, it has a greater value. Notice the arrow at the right end of the number line below. This shows that the whole numbers "keep on keeping on."

Whole Number Line

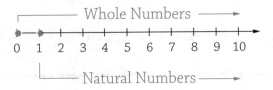

More than likely, you've been introduced to the idea of **negative numbers**, numbers with values less than zero, by now. For example, you may hear about or experienced temperatures "below 0°." They are represented by negative numbers such as −5°F or −3°C. Or, for example, when someone overdraws a checking account, the result is a negative balance such as −$50 or −$100.

It's convenient to think of numbers as positive *or* negative. The counting numbers are all positive, but we usually don't put plus signs (+) in front of them. The fact that they are positive is simply assumed. On the other hand, we always use negative signs (−) when writing negative numbers. The number 0 is a special case. Zero is considered neither positive nor negative. Taken together all positive numbers, zero, and all negative numbers form the set of numbers called **integers**.

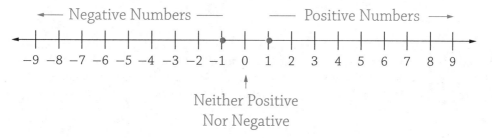

Any number that can be expressed in the form of a ratio between two numbers is part of the set of **rational numbers**. Rational numbers are written as $\frac{a}{b}$, where a and b are integers. There's another extremely important idea that must be added to this definition: $b \neq 0$, that is, b can never equal 0! When you think about it, this shouldn't come as a big surprise. When you first studied division, you learned that dividing by 0 is never permitted in math. That explains why b never equals 0.

Rational numbers include not only fractions such as $\frac{2}{5}$ and $\frac{5}{12}$. Whole numbers are also rational numbers. Why? Simply stated, any whole number can be written as a fraction by placing it as the numerator over the denominator 1 or −1. Therefore, 3 equals $\frac{3}{1}$ and −5 equals $\frac{-5}{1}$ or $\frac{5}{-1}$. Although 0 can never be a denominator, it can be a numerator: $\frac{0}{5} = 0$ and $\frac{0}{-5} = 0$. In fact, 0 divided by any number has a quotient of 0! Here's an example of what one rational number line looks like.

Rational Number Line

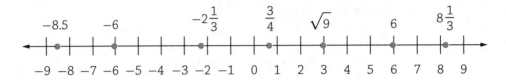

 IRL The idea that rational numbers are "infinitely dense" goes back more than 2,500 years to ancient Greece. Although this discovery is usually attributed to the Pythagorean mathematicians, Chinese and Indian mathematicians had discovered the same concept hundreds of years earlier.

EXAMPLE

▶ Where on the below number line would you mark points that represent $\frac{7}{2}$ and $\frac{-3}{2}$?

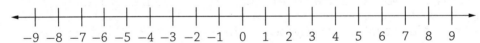

▶ Name the intervals with a denominator of 2 from $\frac{-5}{2}$ to $\frac{5}{2}$. Locate and mark the point on the number line that is midway between $\frac{6}{2}$ and $\frac{8}{2}$. Then, locate and mark the point on the number line that is midway between $\frac{-2}{2}$ and $\frac{-4}{2}$.

▶ $\frac{7}{2}$ and $\frac{-3}{2}$ are located at the points shown on the number line below.

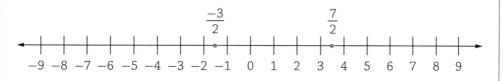

Some numbers cannot be expressed as a simple fraction, so by definition they are not a rational numbers. What are these strange numbers called? **Irrational numbers** are numbers that cannot be written in the form $\frac{a}{b}$, where $b \neq 0$. Notice that the term *irrational* has nothing to do with the idea of being "unreasonable" or "illogical"!

Another way to think about rational numbers is to consider the square roots of whole numbers. Recall that whole numbers include 0 and all the positive integers. Consider the following set of rational numbers:

$N = \{1, 2, 3, \mathbf{4}, 5, 6, 7, 8, \mathbf{9}, 10, 11, 12, 13, 14, 15, \mathbf{16}, 17, 18, 19, 20, 21, 22, 23, 24, \mathbf{25}\}$

Notice that the numbers 4, 9, 16, and 25 appear in bold type. Each of these numbers is a **perfect square**, a number that can be expressed as the product of two equal integers. For example, $4 = 2 \times 2 = 2^2$ and $4 = -2 \times -2 = 2^2$.

> **BTW**
> Probably the most famous irrational number is **pi** (π), the ratio of the circumference of a circle to its diameter. When we work with pi in math, we use $\frac{22}{7}$ or 3.14 to do calculations. However, don't let these values fool you! These are just approximations of pi's value.

All the other bold numbers listed above also satisfy the definition of a perfect square:

$$9 = 3 \times 3 = 3^2 \quad \text{and} \quad 9 = -3 \times -3 = 3^2$$
$$16 = 4 \times 4 = 4^2 \quad \text{and} \quad 16 = -4 \times -4 = 4^2$$
$$25 = 5 \times 5 = 5^2 \quad \text{and} \quad 25 = -5 \times -5 = 5^2$$

If you look at a 10-by-10 multiplication table, the pattern of perfect squares becomes obvious.

×	1	2	3	4	5	6	7	8	9	10
1	1	2	3	4	5	6	7	8	9	10
2	2	4	6	8	10	12	14	16	18	20
3	3	6	9	12	15	18	21	24	27	30
4	4	8	12	16	20	24	28	32	36	40
5	5	10	15	20	25	30	35	40	45	50
6	6	12	18	24	30	36	42	48	54	60
7	7	14	21	28	35	42	49	56	63	70
8	8	16	24	32	40	48	56	64	72	80
9	9	18	27	36	45	54	63	72	81	90
10	10	20	30	40	50	60	70	80	90	100

All the perfect squares fall on a diagonal from the least to the greatest number. If you extended the grid to 20 by 20, or 50 by 50, 100 by 100, and so on, the pattern continues: All the perfect squares fall on the diagonal.

What does this have to do with irrational numbers? Well, if you take the square root of any whole numbers that is not a perfect square, the result is an irrational number. For example, $\sqrt{2}$ is approximately 1.4142135623723095...; √3 is 1.732050807568877...; and √5 is 2.23606797749979.... Another way to say this is that the decimal representations of all irrational numbers are nonending and nonrepeating! When taken together, the set of all rational numbers and irrational numbers form the **real numbers**.

Real Number Line

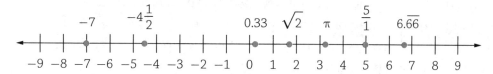

BTW

When you create a number line to solve a problem, you only need to label a few intervals and points. There's rarely any practical purpose in drawing "perfect" representations of the number line when just the essentials will do!

EXAMPLE

▶ Where on this number line would you mark points that represent −5 and √8?

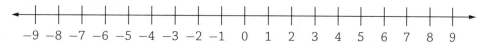

▶ Locate and mark the point on the number line that is midway between −4 and −6. Then, use a calculator to find and locate the point on the number line that represents √8.

▶ −5 and √8 are located at the positions shown on the number line below.

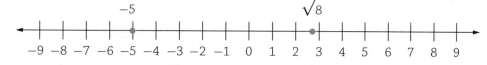

Using a Number Line

Number lines are helpful when you are comparing the value of numbers. For example, you can use a number line to compare two numbers such as $\frac{-3}{7}$ and $\frac{-1}{2}$ to determine which is greater than the other.

The first thing to notice is that the two numbers have different denominators. You can only compare the numbers by finding a common denominator. Since $2 \times 7 = 14$, you must convert both numbers to a rational number with a denominator of 14:

$$\frac{-3}{7} \times \frac{2}{2} = \frac{-6}{14}$$

$$\frac{1}{2} \times \frac{7}{7} = \frac{-7}{14}$$

Next, you must mark the two numbers on a number line to make the comparison easy to see. Then, you should identify which number is greater by noting whether it is to the right or left of the other number. Since $\frac{-6}{14}$ is to the right of $\frac{-7}{14}$, $\frac{-3}{7}$ is greater than $\frac{-1}{2}$:

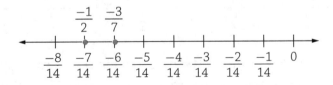

> **EXAMPLE**
>
> ▶ Which number is smaller: $\frac{-3}{5}$ or $\frac{-4}{9}$? Use a number line to compare the numbers.
>
> ▶ Express both numbers so they have a common denominator.
>
> $$\frac{-3}{5} \times \frac{9}{9} = \frac{-27}{45}$$

$$\frac{-4}{9} \times \frac{5}{5} = \frac{-20}{45}$$

▶ Mark the numbers on a number line.

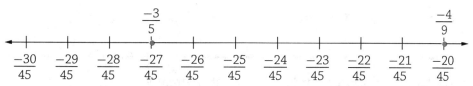

▶ $\frac{-3}{5}$ is less than $\frac{-4}{9}$ since it is to left of $\frac{-4}{9}$ on the number line.

Number lines are very helpful when you want to order numbers from least to greatest or from greatest to least. For example, suppose you are asked to order the numbers $\frac{-7}{9}$, -0.25, 0.15, and $\frac{5}{8}$.

Notice that two of the numbers are written as decimals and two as fractions. The first thing to do is to make sure that all the numbers are expressed in the same form. It's easy to find the decimal value of a fraction—all you have to do is divide the numerator by the denominator. So, $\frac{-7}{9} \approx -0.777$ and $\frac{5}{8} = 0.623$.

Now, you must mark the positions of the four numbers and compare their locations on a number line. Remember: Moving from left to right, the numbers get larger; and moving right to left, the numbers get smaller.

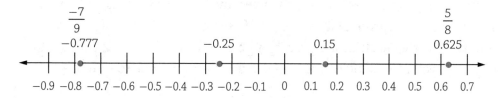

Moving from left to right, it's easy to see that the numbers ordered from least to greatest are: $\frac{-7}{9}$, -0.25, 0.15, and $\frac{5}{8}$. Moving from right to left, the numbers are ordered from greatest to least: $\frac{5}{8}$, 0.15, -0.25, and $\frac{-7}{9}$.

> **EXAMPLE**
>
> ▶ Place these numbers in order from least to greatest: $\frac{-2}{5}$, $\frac{1}{5}$, 0.45, -0.35. Use a number line.
>
> ▶ Express all numbers so they have decimal form.
>
> $$\frac{-2}{3} \approx -0.67$$
>
> $$\frac{1}{5} = 0.20$$
>
> ▶ Mark the numbers on a number line.
>
>
>
> ▶ In order from least to greatest the numbers are $\frac{-2}{3}$, -0.35, $\frac{1}{5}$, and 0.45.

You're already familiar with the principles involved in rounding. When you round a number to a given place value, you first must look at the digit to the right of the place value. If that digit is 5 or greater, you round up. If that digit is 4 or smaller, you round down. You can also use a number line to round.

CHAPTER 1 The Number System

> **EXAMPLE**
>
> ▶ What is 4.52 rounded to the nearest tenth? Use a number line.
>
> ▶ Create a number line to represent the numbers in the problem. Then mark the number you are rounding.
>
>
>
> ▶ Determine which decimal your number is closer to: 4.52 is closer to 4.5 than it is to 4.6.
>
> ▶ 4.52 rounded to the nearest tenth is 4.5.

Absolute Value

You know that on a number line, 4 is located four places to the right of 0. On the other hand, −4 is four places to the left of 0. If you put the direction and signs aside, both 4 and −4 are the same *distance* from 0, as you can see on the number line below.

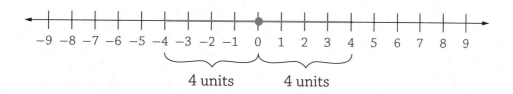

The absolute value of a number is its distance from 0 on the number line. We write the absolute value of a number by placing it between vertical bars—for example, $|4| = 4$ and $|-4| = 4$.

> **EXAMPLE**
>
> ▶ What is the absolute value of $|-23|$ and $|23|$? Explain what absolute value means in relation to these numbers.

> Write the number between the two vertical bars without signs.
>
> The absolute value of |−23| and |23| is 23. Each number is 23 units from 0.

Performing Operations with Whole Numbers

Addition involves finding the sum of two or more numbers called **addends**. No doubt, you're quite familiar with the process of addition. The key to accuracy is lining up the digits according to place value—all the ones in the first column at the right, the tens in the second column from the right, and so on. Then, you must add together the numbers that share the same place. If the sum of a column is greater than 10, you must regroup and carry the 1 to the next column to the left.

You can model addition using base-10 "blocks." For example, suppose you are asked to find the sum of 24 and 38. Here's what a model would look like:

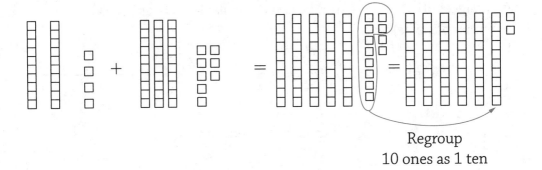

Regroup
10 ones as 1 ten

The standard way to write and solve this problem is:

$$\begin{array}{r} 1 \\ 24 \\ +\ 38 \\ \hline 62 \end{array}$$

CHAPTER 1   The Number System

Subtraction is the opposite of addition since it involves taking away one number from another. The standard way to write and solve a subtraction problem is:

$$
\begin{array}{r} 56 \\ -29 \\ \hline \end{array} \quad \Rightarrow \quad \begin{array}{r} {}^{4\,1}\!\!\not{5}6 \\ -29 \\ \hline 27 \end{array}
$$

EXAMPLE

▶ What is the difference between 38 and 9?

▶ Write the problem in vertical format.

$$
\begin{array}{r} 38 \\ -9 \\ \hline \end{array}
$$

▶ Change 1 ten into 10 ones.

$$
\begin{array}{r} {}^{2\,1}\!\!\not{3}8 \\ -9 \\ \hline 29 \end{array}
$$

Multiplication can be thought of as repeated addition. When we say, "12 × 5," what we mean is "12 + 12 + 12 + 12 + 12." The standard way to write and solve this problem is shown below. We carry a 1 into the tens spot when we multiply 2 and 5:

$$
\begin{array}{r} {}^{1} \\ 12 \\ \times5 \\ \hline 0 \end{array} \quad \Rightarrow \quad \begin{array}{r} {}^{1} \\ 12 \\ \times5 \\ \hline 60 \end{array}
$$

> **EXAMPLE**
>
> ▸ What is the product of 14 and 6?
>
> ▸ Write the problem in vertical format.
>
> $$\begin{array}{r} 14 \\ \times\ 6 \\ \hline 84 \end{array}$$
>
> ▸ Since 6 times 4 equals 24, write the 4 in the ones place and carry the 2 to the tens place.
>
> $$\begin{array}{r} {}^{2}\ \ \\ 14 \\ \times\ 6 \\ \hline 4 \end{array}$$
>
> ▸ Complete the multiplication: 6 times 10 is 60, and add the 20 carried over to make 80:
>
> $$\begin{array}{r} {}^{2}\ \ \\ 14 \\ \times\ 6 \\ \hline 84 \end{array}$$
>
> ▸ The product of 14 times 6 is 84.

Since multiplication can be thought of as repeated addition, it's not surprising that division can be thought of as repeated subtraction. When we say 12 divided by 4, what we mean is:

$$\begin{array}{r} 12 \\ -\ 4 \\ \hline 8 \\ -\ 4 \\ \hline 4 \\ -\ 4 \\ \hline 0 \end{array}$$

So, 12 divided by 4 equals 3; that is, 12 can be "divided" into 3 equal groups of 4.

When you divide greater numbers, it's easier to use long division. The number you are dividing is called the **dividend**, and the number you are dividing by is called the **divisor**. For example, in the problem 123 divided by 8, the number 123 is the dividend and the number 8 is the divisor. You can use long division to solve the problem:

$$
\begin{array}{r}
15\ \text{R}3 \\
8\overline{)123} \\
-\underline{8}\downarrow \\
43 \\
-\underline{40} \\
3
\end{array}
$$

The answer in a division problem is called the **quotient**. In this problem, the quotient 15 R 3 means that 8 goes into 123 evenly 15 times, plus 3 ones "remaining," or left over.

EXAMPLE

▸ What is 286 divided by 13?

▸ Solve the problem using long division.

$$
\begin{array}{r}
22 \\
13\overline{)286} \\
-\underline{26}\downarrow \\
26 \\
-\underline{26} \\
0
\end{array}
$$

▸ 286 divided by 13 is 22.

You can see how addition and subtraction are related by creating **fact families**, that is, sentences using the same numbers. For example:

$$8 + 4 = 12 \qquad 12 + 9 = 21$$
$$4 + 8 = 12 \qquad 9 + 12 = 21$$
$$12 - 4 = 8 \qquad 21 - 9 = 12$$
$$12 - 8 = 4 \qquad 21 - 12 = 9$$

> ▶ What addition and subtraction fact family can be created using 5, 9, and 14?
>
> ▶ Write two addition sentences using the numbers.
>
> $$5 + 9 = 14 \qquad 9 + 5 = 14$$
>
> ▶ Write two subtraction sentences using the numbers.
>
> $$14 - 5 = 9 \qquad 14 - 9 = 5$$
>
> ▶ 5, 9, and 14 form the following fact family:
>
> $$5 + 9 = 14 \qquad 9 + 5 = 14 \qquad 14 - 5 = 9 \qquad 14 - 9 = 5$$

Notice how the same three numbers are used to form a fact family that consists of two addition and two subtraction sentences. Similarly, three related numbers form multiplication and division fact families:

$$3 \times 8 = 24 \qquad 5 \times 9 = 45$$
$$8 \times 3 = 24 \qquad 9 \times 5 = 45$$
$$24 \div 8 = 3 \qquad 45 \div 9 = 5$$
$$24 \div 3 = 8 \qquad 45 \div 5 = 9$$

EXAMPLE

▶ What multiplication and division fact family can be created using 6, 7, and 42?

▶ Write two multiplication sentences using the numbers.

$6 \times 7 = 42$ $\qquad$ $7 \times 6 = 42$

▶ Write two division sentences using the numbers.

$42 \div 6 = 7$ $\qquad$ $42 \div 7 = 6$

▶ 6, 7, and 42 form the following fact family:

$6 \times 7 = 42$ $\qquad$ $7 \times 6 = 42$ $\qquad$ $42 \div 6 = 7$ $\qquad$ $42 \div 7 = 6$

Fact families show how addition and subtraction are **inverse**, or opposite, operations that undo each other. Fact families also show how multiplication and division are inverse operations.

Properties of Numbers

When you perform any one of the four basic operations on numbers, it's essential that you understand how the numbers work together. Some operations have certain properties because they work in established ways. For example, when you add two or more numbers, does it matter the order you add in? Think about adding 12, 9, and 18:

```
    9         9        18        18
   12        18         9        12
 + 18      + 12      + 12      +  9
 ----      ----      ----      ----
   39        39        39        39
```

It's obvious that the order you add numbers in doesn't change the sum. The same is true of multiplication. Multiplying 11 times 37 will result in the same product as 37 times 11. This fact is called the **commutative property of addition and multiplication.** Although addition and multiplication are commutative, subtraction and division are *not*. The order of numbers in a subtraction or a division problem matters in the difference or quotient. The difference between 56 and 24 is not the same as 24 minus 56, and 72 divided by 6 is not the same as 6 divided by 72.

$$\begin{array}{r} 56 \\ -24 \\ \hline 32 \end{array} \qquad \begin{array}{r} 24 \\ -56 \\ \hline -32 \end{array} \qquad 72 \div 6 = 12 \qquad 6 \div 72 = 0.083$$

Addition and multiplication share another property. The **associative property of addition** states that the value of the sum of two or more numbers does not depend on how they are grouped:

$$(12 + 7) + 6 = 25 \qquad 12 + (7 + 6) = 25$$

In general terms, the associative property of addition states:

$$(a + b) + c = a + (b + c)$$

Likewise, the **associative property of multiplication** states that the value of the product of two or more numbers does not depend on how the numbers are grouped:

$$(5 \times 8) \times 3 = 120 \qquad 5 \times (8 \times 3) = 120$$

In other words, the associative property of multiplication states:

$$(a \times b) \times c = a \times (b \times c)$$

EXAMPLE

▶ Which property do the following examples show?

$$8 + 7 + 3 = 18 \qquad 6 \times 8 = 48$$

$$3 + 7 + 8 = 18 \qquad 8 \times 6 = 48$$

- Notice that the addends and the sums of the addition equations are the same. The factors and product in the multiplication equations are the same. Only the order changes.

- The examples show the **commutative property of addition and multiplication**.

There are some related properties, called **identity properties**, that are connected to addition and multiplication. For example, the sum of any number and 0 is that number. The product of any number and 0 is always 0, and the product of any number and 1 is always that number. No doubt, you are familiar with using these properties even if you don't know their official names!

$$12 + 0 = 12 \qquad 12 \times 0 = 0 \qquad 12 \times 1 = 12$$

The distributive property of multiplication over addition states that multiplying a number and a sum is the same as multiplying the number and each part of the sum and then adding the products. For example:

$$7 \times (5 + 8) = 7 \times 5 + 7 \times 8 = 35 + 56 = 91$$

In general terms, the **distributive property of multiplication over addition** states that $a(b + c) = ab + ac$.

There are two other properties, called **inverse properties**, that are important when adding and multiplying. The **inverse property of addition** states that for any number, adding its inverse, or opposite value, results in a sum of 0. For example, $5 - 5 = 0$ and $123{,}456 - 123{,}456 = 0$.

In a similar way, the **inverse property of multiplication** states that multiplying any number by its inverse results in a product of 1. The reason is that multiplying a number by its inverse results in a numerator and denominator that are the same. For example:

$$5 \times \frac{1}{5} = \frac{5}{5} = 1 \qquad\qquad 14{,}649 \times \frac{1}{14{,}649} = \frac{14{,}649}{14{,}649} = 1$$

EXAMPLE

▶ Which property do the following examples show?

$$5(3 + 6) = (5 \times 3) + (5 \times 6)$$

$$12(4 + 7) = (12 \times 4) + (12 \times 7)$$

▶ Notice that the first factor is used to multiply each of the addends in parentheses and then the products are added.

▶ The examples show the distributative property of multiplication over addition.

CHAPTER 1 The Number System

EXERCISES

EXERCISE 1-1

Identify the value of the indicated digit.

1. What is the value of 4 in the number 1,643.527?

2. What is the value of 6 in the number 21,873.164?

EXERCISE 1-2

Mark the indicated points on a number line.

1. Where on this number line would you mark points that represent $\frac{5}{2}$ and $\frac{-7}{2}$?

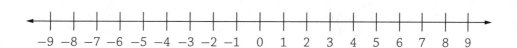

2. Where on this number line would you mark points that represent -3 and $\sqrt{12}$?

EXERCISE 1-3

Use a number line to compare the indicated numbers.

1. Which number is less: $\frac{-5}{7}$ or $\frac{-4}{5}$? Draw a number line to compare the numbers.

2. Draw a number line and place these numbers in order from least to greatest on it: $\dfrac{-2}{3}, \dfrac{1}{5}, 0.45, -0.35$.

EXERCISE 1-4

Round each number to the indicated place value.

1. What is 8.257 rounded to the nearest hundredth?

2. What is 13.382 rounded to the nearest tenth?

EXERCISE 1-5

Identify the absolute value of the indicated number.

1. Mark the absolute value of −5 on the number line.

2. What is the absolute value of $|-18|$ and $|18|$?

EXERCISE 1-6

Interpret the base-10 "blocks" shown to solve each problem.

1. What number is shown below?

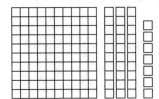

2. What is the sum of the two groups of blocks below?

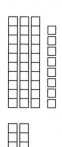

EXERCISE 1-7

Find the sum or difference using a vertical format.

1. Use a vertical format to find the sum of the numbers below.

   ```
     173
      84
   + 211
   ```

2. Use a vertical format to find the difference between the numbers.

   ```
     326
   - 142
   ```

EXERCISE 1-8

Find the product or quotient.

1. What is the product of 16 and 7?

2. What is the quotient of 153 and 9?

EXERCISE 1-9

Identify the sentences that form the indicated fact family.

1. Write an addition and subtraction fact family that uses the numbers 6, 7, and 13.

2. Write a multiplication and division fact family that uses the numbers 7, 12, and 84.

EXERCISE 1-10

Identify the property of numbers shown in the examples.

1. Which property do the following examples show?

   $6 + 8 + 11 = 25 \quad 8 + 11 + 6 = 25$

2. Which property do the following examples show?

   $(9 \times 5) \times 2 = 90 \quad 9 \times (5 \times 2) = 90$

3. Which property does the following example show?

   $5 \times (6 + 8) = (5 \times 6) + (5 \times 8)$

4. Which property does the following example show?

   $9 \times \dfrac{1}{9} = 1$

Flashcard App

Decimals

MUST KNOW

- A decimal is a special kind of fraction whose denominator is a power of ten.

- Decimal place values are called *tenths*, *hundredths*, *thousandths*, and so on.

- Decimals can be compared and put in order using number lines.

- We perform basic operations with decimals the same way as we do with whole numbers, except for the placement of the decimal point.

Working with decimals is a common occurrence in daily life. Just think about how many times each day you deal with money. If you have a $5 bill and spend $1.25, you are dealing with two decimals: $5.00 and $1.25. If you stand on a digital scale to find your weight in pounds, you might see a number such as 83.5. If you look at the odometer the next time you're in a car, you'll see that the car's mileage is shown in decimal form. Can you think of other situations in which decimals are used?

Understanding Decimals

A **decimal** is a special type of fraction, one whose denominator is 10 or a power of 10. In fact, all numbers are decimals, since even whole numbers such as 4 or 121 can be written as 4.0 or 121.0. When a decimal such as 126.35 is written, the **decimal point** (.) separates the whole numbers from the fractional part:

Thousands	Hundreds	Tens	Ones	Tenths	Hundredths	Thousands
	1	2	6 .	3	5	

The number 126.35 means 126 and 35 hundredths. The number 1 is in the hundreds place, so it represents one 100. The number 2 is in the tens place, so it represents two 10s. The number 6 is in the ones place, so it stands for six 1s. The number 3 to the right of the decimal point is in the tenths place and represents 3 tenths. Likewise, the number 5 is in the hundredths place and represents 5 hundredths.

EXAMPLE

▶ Write 273.78 using a place value chart:

Thousands	Hundreds	Tens	Ones	Tenths	Hundredths	Thousands

CHAPTER 2 Decimals **29**

▸ Write each digit so it correctly represents its place value in the decimal.

▸ 273.78 written in a place value chart is:

Thousands	Hundreds	Tens	Ones	Tenths	Hundredths	Thousands
	2	7	3 .	7	8	

Another way to visualize a decimal is to use a hundreds grid. The example below shows 0.57; that is, 57 of the 100 squares are shaded.

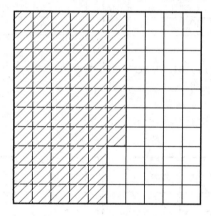

A number such as 2.23 using hundreds grids would look like this:

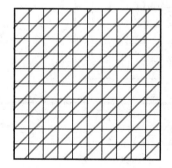 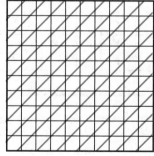 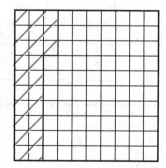

Notice that the 2 in the ones digit is represented by two completely shaded squares.

EXAMPLE

▶ Show 1.45 by correctly shading the hundreds grids.

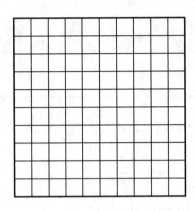 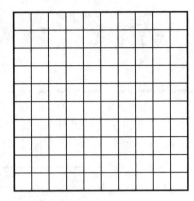

▶ Each small square in a hundreds grid represents 1 one hundredth. The entire grid represents 1.

▶ Using hundreds grids, 1.45 is:

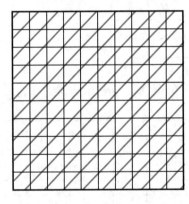 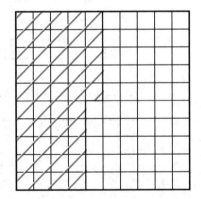

Reading and Writing Decimals

There are several ways to express a decimal. The **word form** of a decimal involves expressions such as "one-tenth," "three hundredths," or "five thousandths." If you look at the columns in the place value chart, you'll notice that the names of the places to the right of the decimal point end in -*th*. There's a big difference between ten and tenth; "ten" is 100 times larger "one-tenth"!

> **EXAMPLE**
>
> ▶ What is the correct word form of 5.314?
>
> ▶ Identify the value of the whole number part of the decimal. Then, identify the place value of the fractional part to the right of the decimal point. The 4 in 314 represents 4 thousandths.
>
> ▶ The correct word form of 5.314 is "five and three hundred fourteen thousandths."

You're already familiar with the **decimal form** of a number. For example, "three and four tenths" is written in decimal form as 3.4, and a number such as "seventy-five hundreds" is written in decimal form as 0.75. Notice that a zero is placed before the decimal point when its value is less than 1. When you translate from word to decimal form, pay attention to the endings of the words. Words that end with the suffix —*th* mean their number representation comes after the decimal point.

> **EXAMPLE**
>
> ▶ What is the decimal form of "sixteen and twenty-one hundredths"?
>
> ▶ Identify the value of the whole number part of the decimal. Then, identify how many places to the right of the decimal point represents the place value of the fractional part. "Twenty-one hundreds" is two places to the right of the decimal point.
>
> ▶ The correct decimal form of "sixteen and twenty-one hundredths" is 16.21.

A third way to express a decimal is to write its **expanded form**. In the expanded form of a decimal, each digit is expressed based on its place value. Remember that a decimal such as 215.36 can be thought of as 2 one hundreds plus 1 ten plus 5 ones plus 3 tenths plus 6 hundredths. In expanded form, you would write the number as:

$$2 \times 100 + 1 \times 10 + 5 \times 1 + 3 \times \frac{1}{10} + 6 \times \frac{1}{100}$$

If a number has a 0, just leave out that place value. For example, the expanded form of 103.405 is written as:

$$1 \times 100 + 3 \times 1 + 4 \times \frac{1}{10} + 5 \times \frac{1}{1000}$$

Notice that the expanded form of this number has zeros in the tens place and hundredths place. Therefore, its expanded form does not include representations of these places.

BTW
Mastering the writing of decimals in expanded form will sharpen your understanding of place value. You'll also apply this skill when you learn to write numbers using scientific notation.

EXAMPLE

▶ How do you write 137.25 in expanded form?

▶ Identify the value of each digit moving from left to right based on its place. Write the value of the digit using place value. The 1 represents 1×100. The 3 represents 3×10. The 7 represents 7×1. The 2 in 137.25 represents $2 \times \frac{1}{10}$ and the 5 represents $5 \times \frac{1}{100}$.

▶ 137.25 written in expanded form is:

$$1 \times 100 + 3 \times 10 + 7 \times 1 + 2 \times \frac{1}{10} + 5 \times \frac{1}{100}$$

Comparing and Ordering Decimals

Sometimes, we want to compare the value of two or more decimals. Locating the decimals on a number line is a foolproof method. For example, suppose we want to compare and order the decimals 1.53, 1.79, 1.51, and 1.66. The first thing to do is create a number line that covers the range of values we will deal with. In this case, a number line that starts at 1.5 and ends at 1.8 does the trick because all the decimals being ordered fall within the range of the least and greatest number.

Next, we need to locate each decimal on the number line and mark its location, making sure to write the value we are marking above each decimal. When we're finished, the number line will look like this:

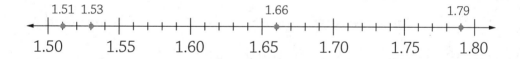

In order from least to greatest value, the decimals are 1.51, 1.53, 1.66, and 1.79. In order from greatest to least, the decimals are 1.79, 1.66, 1.53, and 1.51. Notice that we can use our knowledge of how number lines work to compare decimals. For example, we can tell that 1.53 is greater than 1.51 because 1.53 is to the right of 1.51. Similarly, 1.51 is less than 1.53 because 1.51 is to the left of 1.53.

A handy way to record our comparison and ordering of decimals is by using the symbols for greater than (>) and less than (<):

$1.79 > 1.66 > 1.53 > 1.51$ greatest to least
$1.51 < 1.53 < 1.66 < 1.79$ least to greatest

Note that two decimals such as 1.53 and 1.530 are actually equal in value. They're just written to different decimal places.

Rounding Decimals

Some problems we encounter in math class and in the real world ask us to round a decimal to a particular place value. For example, what is the sum of 2.32 and 4.51 rounded to the nearest tenth? Rounding decimals involves the same principles as rounding whole numbers. So, $2.32 + 4.51 = 6.83$ and rounded to the nearest tenth is 6.8. We need to be careful to note the digit we are rounding to since it may change the answer. For example, the sum of 3.48 and 5.19 rounded to the nearest tenth is 8.7, not 8.6. Here's why: $3.48 + 5.19 = 8.67$.

However, the sum 8.67 is written to the nearest hundredths. To find the sum rounded to the nearest tenth, we must round the hundredths place. In this example, $7 > 5$, so we must round 8.67 up to 8.7.

> **EXAMPLE**
>
> ▶ What is the difference between 13.765 and 2.121 to the nearest tenth?
>
> ▶ Find the difference between the two numbers: $13.765 - 2.121 = 11.644$. Identify the digit that represents the place value you are finding. Then, check the digit to right of this place. Round up the digit you are looking for if the digit to its right is 5 or greater. Keep the digit you are looking for the same if the digit to its right is less than 5.
>
> ▶ The difference between 13.765 and 2.121 to the nearest tenth is is 11.6.

Sometimes a decimal is very small—for example, 0.0005684—and the best way to work with it is to round the decimal to its leading digit. The **leading digit** is the first digit of the decimal that is not zero. In 0.0005734, the leading digit is 5. To round this decimal to its leading digit, we must look to the digit immediately to the right of 5. In the example, this is 7. Now, since 7 is greater than 5, we would round the leading digit up. Therefore, 0.0005734 rounded to its leading digit is 0.0006.

> **EXAMPLE**
>
> ▶ What is 0.00074392 rounded to its leading digit?
>
> ▶ Identify the leading digit. Then check the digit to right of the leading digit. Round the leading digit up if the digit to its right is 5 or greater. Keep the leading digit the same if the digit to its right is less than 5.
>
> ▶ 0.00074392 rounded to its leading digit is 0.0007.

Adding and Subtracting Decimals

Adding and subtracting decimals uses the same set of skills we use when adding and subtracting whole numbers. The only significant way these operations differ is the need to place the decimal point.

As with whole numbers, we can use hundred grids to model the addition of two decimals. For example, what is the sum of 0.35 + 0.23? First, we mark 35 of 100 small squares on the grid to represent 0.35. Then, using the same grid, we mark an additional 23 small squares to represent 0.23:

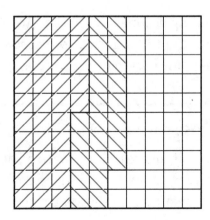

Now, all we have to do is count the number of small squares that are marked after representing both numbers. Since each column represents 10 hundredths, we can count the number of squares quickly: $0.35 + 0.23 = 0.58$.

EXAMPLE

▶ Use a hundreds grid to find the sum of 0.41 and 0.33.

▶ Mark 41 of the 100 small squares. Then, mark another 33 of the small squares.

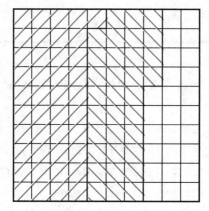

▶ Count the total number of marked small squares: $0.41 + 0.33 = 0.74$.

How do we handle adding decimals when they are greater than 1 or when their sum is greater than 1? In fact, all we need to do is represent the numbers using more than one hundreds grid. It's as simple as that!

EXAMPLE

▶ Use a hundreds grid to find the sum of 0.72 and 0.55.

▶ Mark 72 of the 100 small squares. Next, mark 55 additional small squares. Since there aren't enough small squares unmarked on the original grid, continue marking on a second grid.

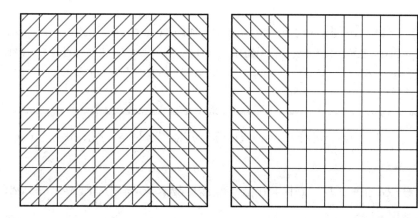

▶ Count the total number of marked small squares. Remember that a completely marked hundreds grid equals 1: $0.72 + 0.55 = 1.27$.

Subtracting two decimals is just the opposite of adding them. Start by representing the greater number. Mark the appropriate number of small squares on one or more hundreds grids. Then, cross out the number of small squares based on the lesser number. What's left over is the difference. For example, we would show 0.87 minus 0.45 as:

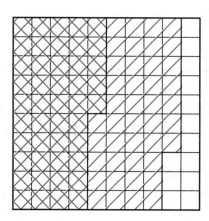

When we are finished representing both numbers in the subtraction problem, count the number of small squares that are still marked. In this case, there are 42 small squares still marked, so $0.87 - 0.45 = 0.42$.

EXAMPLE

▸ What is the difference between 0.92 and 0.35? Use a hundreds grid to solve.

▸ Mark 92 of the 100 small squares. Next, cross out 35 of the marked small squares.

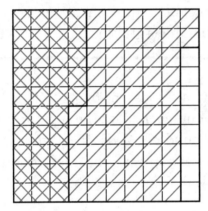

▸ Count the total number of marked small squares that have not been crossed out: $0.92 - 0.35 = 0.57$.

We need to use two or more hundreds grids to subtract decimals greater than 1, just as we did when we added two decimals greater than 1.

EXAMPLE

▸ What is the difference between 1.45 and 0.75? Use hundreds grids to solve.

▸ Represent 1.45 on hundreds grids. Remember that a completely marked hundreds grid equals 1. Next, cross out 75 of the marked small squares.

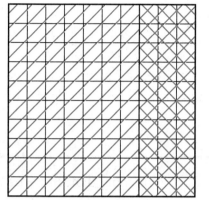 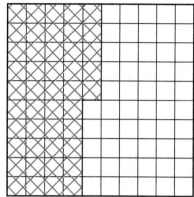

▶ Count the total number of marked small squares that have not been crossed out: $1.45 - 0.75 = 0.70$, or 0.7.

We can also use a vertical format to add and subtract decimals, just as we do with whole numbers. The first thing to do is line up our numbers along the decimal points. This is very important since it helps guarantee that we are adding or subtracting numbers with the same place value. Then, we should add zeros as placeholders if the number of digits in our decimals is not equal. After that, we simply add or subtract the numbers in each column. Here is an example of how to use a vertical format to add two decimals.

BTW

Blank number lines, place value charts, and hundreds grids that you can easily customize are found on many math and science Internet sites.

EXAMPLE

▶ Find the sum of 4.038 and 11.56.

▶
$$
\begin{array}{r}
4.038 \\
+\ 11.560 \\
\hline
15.598
\end{array}
$$

Notice that 11.56 is represented as 11.560 in order to keep the number of digits in both decimals the same. Therefore, the sum of 4.038 + 11.56 is 15.598. The same vertical format can be used to add three or more decimals. Just remember to keep the number of places in each decimal the same.

We use the same principles when using a vertical format to subtract two decimals.

EXAMPLE

▶ Find the difference between 6.538 and 3.13. Remember that, since the decimal 6.538 goes to the thousandths place, you must add a zero to 3.13.

▶
```
   6.538
 − 3.130
   3.408
```

Let's try another.

EXAMPLE

▶ What is the sum of 5.412 + 2.16? Solve the problem using a vertical format.

▶ Write the decimals so that the place value of the digits lines up based on the decimal points. Add zeros where they are needed.

```
   5.412
 + 2.160
   7.572
```

When we are subtracting whole numbers and decimals, we may need to add more than one zero.

EXAMPLE

▶ What is the difference between 7 and 3.15? Solve the problem using a vertical format.

▶ Write the decimals so that the place value of the digits line up based on the decimal points. Add zeros where they are needed.

$$\begin{array}{r} 7.00 \\ -\ 3.15 \\ \hline 3.85 \end{array}$$

Some problems don't require us to find an exact answer—a good estimate will do. For example, suppose you earned $30 for walking your neighbor's dog while she was on vacation. You want to buy three books that cost $9.75, $10.99, and $7.95. Will you have enough money to buy the books?

The simplest way to answer your question is to use rounding to estimate the sum. In this case, since you want to be sure you have enough money, you decided to round up each price to the nearest dollar:

$9.75 → $10

$10.99 → $11

$7.95 → $8

Find the sum of 10, 11, and 8 using mental math. Since you rounded up and the sum is $29, you'll have enough for all three books!

EXAMPLE

▶ You are going on vacation to a beach resort. You want to buy a new T-shirt for $12.95, new shorts for $16.50, and new swimming trunks for $15.75. Use mental math to determine if $40 is enough to make your purchases.

▶ Round each price up to the nearest dollar value.

$12.95 → $13

$16.50 → $17

$15.75 → $16

- Find the sum of the rounded numbers: 13 + 17 + 16 = (13 + 17) + 16 = 30 + 16 = 46.

- The estimated sum of the three items is $46.00. $40 is not enough money to purchase them.

Multiplying Decimals

We can use a hundreds grid to model the multiplication of two decimals by following the same procedure we used for modeling the multiplication of whole numbers. For example, suppose we want to model 0.6 times 0.8. We would mark the grid 6 squares down and 8 squares across as shown below. By counting the marked squares, we can readily see that the product of 0.6 and 0.8 equals 0.48, or "forty-eight hundredths":

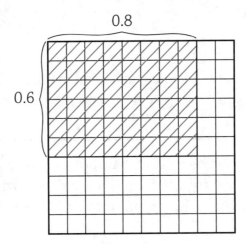

EXAMPLE

- What is the product of 0.5 and 1.2? Model your solution using hundreds grids.

▶ Notice that one factor is greater than 1. This means that you'll need two hundreds grids to model the problem.

▶ First, you mark 5 small squares down. Then, you mark 12 small square across.

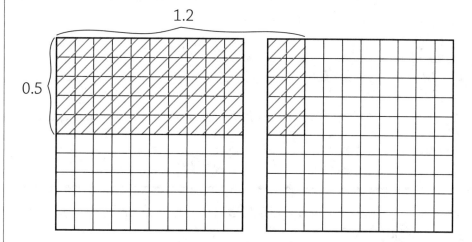

▶ Count the marked squares. The product of 0.5 and 1.2, therefore, is 0.60.

Using hundreds grids to model problems that involve the multiplication of decimals helps us understand what's involved in the operation. However, using models can become very complicated once the factors become greater. Besides, if we know how to multiply whole numbers—and, we do!—working with decimals involves just one extra step. Once we find the product, we must correctly place the decimal point in the product. Here's an example of how to go about doing this.

$$\begin{array}{r} 3.64 \\ \times\ 0.28 \\ \hline 2912 \\ 728 \\ \hline 10192 \end{array} \rightarrow \text{Where does the decimal point go?}$$

To determine where to place the decimal point, we must look at the factors in the problem. Why? The number of decimal places in the product is equal to the sum of the number of decimals places in the factors. The two factors in the problem are 3.65 and 0.28. Each factor has two places after the decimal point. So, the product of the problem will have four decimal places.

```
   3.64
 × 0.28
  2912
   728
 1.0192  ← Count four places from the last digit moving right to left.
```

So, the product of 3.64 × 0.28 is 1.0192.

> **EXAMPLE**
>
> ▶ What is the product of 10.35 and 1.525? Use vertical multiplication to find the solution.
>
> ▶ Find the product as if you are solving a multiplication problem involving whole numbers.
>
> ```
>     10.35
>  × 1.525
>    5175
>    2070
>    5175
>    1035
>  1578375
> ```
>
> ▶ Find the sum of the number of decimal places in both factors. Count five places from the last digit moving right to left and place the decimal point. The product of 10.35 and 1.525, then, is 15.78375.

To solve some problems involving decimals, we may need to multiply and add.

EXAMPLE

▶ At the local farmer's market, Abdul buys 3.5 pounds of potatoes and 4.8 pounds of squash. The red potatoes cost $1.19 a pound, and the squash costs $0.99 a pound. How much did Abdul spend to the nearest penny?

▶ Use vertical multiplication to solve each problem without placing the decimal point.

Red Potatoes	**Squash**
1.19	0.99
× 3.5	× 4.8
595	792
357	396
4165	4752

▶ Find the sum of the decimal places in each problem, and mark where the decimal point goes. Round each product to the nearest cent.

　　Red Potatoes → $4.165 → $4.17

　　Squash → $4.752 → $4.75

▶ Add the costs of both purchases.

　　$4.17 + $4.75 = $8.92

▶ Abdul spent a total of $8.92 for the red potatoes and squash.

Dividing Decimals

We've already seen that multiplying decimals follows the same steps as multiplying whole numbers with the added step of placing the decimal point correctly in the product. Similarly, dividing decimals follows the same steps as long division of whole numbers, except for the placement of the decimal point in the quotient. So, where does the decimal point go? Where we place the decimal point in a division problem depends on where the decimal point appears in the dividend.

Consider this problem. Jane and her friends bought seven tickets for a Saturday matinee of a new action film. The tickets cost a total of $82.25. How much did each ticket cost?

$$
\begin{array}{r}
11.75 \\
7 \overline{)82.25} \\
-7 \\
\hline
12 \\
-7 \\
\hline
52 \\
-49 \\
\hline
35 \\
-35 \\
\hline
0
\end{array}
$$

So, each movie ticket cost $11.75. Notice that the decimal point in the quotient is in the same exact position as the decimal point in the dividend.

EXAMPLE

▶ Melinda bought 11 color pastel pencils for her art lessons. She spent a total of $16.39. How much did each pastel pencil cost?

▶ Write the problem in long division format.

$$11 \overline{)16.39}$$

▶ Solve the division problem. Since 11 goes into 16 once, write a 1 above 16, as shown below, and then a decimal point after the 1.

$$\begin{array}{r} 1.49 \\ 11\overline{)16.39} \\ -11 \\ \hline 53 \\ -44 \\ \hline 99 \\ -99 \\ \hline 0 \end{array}$$

▶ Each pastel pencil cost $1.49.

In the previous example, a decimal was divided by a whole number. In some problems, both the divisor and the dividend are decimals. In examples like these, we may need to add 0s, being especially careful about the placement of the decimal point in the quotient.

EXAMPLE

▶ Mr. Salerno paid $35.34 for 12.4 gallons of gasoline. What is the gasoline's price per gallon?

▶ Write the problem in long division format.

$$12.4\overline{)35.34}$$

▶ Multiply the divisor by a power of ten, so it is a whole number. Think: How many places to the right must the decimal point be moved to make the divisor a whole number? Then, multiply the dividend by the same power of ten. In this problem, multiply both the divisor and dividend by 10.

$$12.4\overline{)35.34}$$

▸ Find the quotient by using long division. Add zeros at the end of the dividend if needed.

```
          2.85
    124 ) 353.40    ← Add a zero.
         -248 ↓
          1054
          -992 ↓
            620
           -620
              0
```

▸ The gasoline Mr. Salerno purchased costs $2.85 per gallon.

EXERCISES

EXERCISE 2-1

Answer each question using the given model.

1. Create a place value chart and write 312.649 on it.

2. Draw a hundreds grid and represent 0.37 on it.

EXERCISE 2-2

Write the correct representation of the number.

1. What is the correct word form of 71.426?

2. What is the correct decimal form of "thirty five and nineteen thousandths"?

3. What is the expanded form of 649.328?

4. What is the expanded form of 0.743?

EXERCISE 2-3

Compare or order the numbers.

1. Which is greater: 0.3164 or 0.3194?

2. Compare 1.69, 1.56, 1.72, and 1.52 by marking them on a number line.

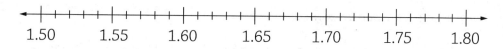

3. Order 6.010, 6.011, 6.001, and 6.101 from greatest to least.

4. Order 3.25, 3.23, 3.33, and 3.22 from least to greatest.

EXERCISE 2-4

Round the numbers to the indicated place value.

1. Round 5.096 to the nearest tenth.

2. Round 2.745 to the nearest hundredth.

3. Round 0.00792 to its leading digit.

4. Round 0.000421 to its leading digit.

EXERCISE 2-5

Solve these addition problems.

1. Use a hundreds grid to find the sum of 0.43 and 0.26.

2. Use a vertical format to show the sum of 5.712 and 12.18.

3. Mikela jogs on Mondays, Wednesdays, and Fridays. This week she jogged 6.437 km on Monday, 5.362 km on Wednesday, and 7.180 km on Friday. How many kilometers did Mikela jog this week?

4. Rashid does volunteer work in his neighborhood. Each week he collects trash on Wednesdays and Saturdays. Last Wednesday, he collected 35.73 kilograms of trash and on Saturday he collected 47.39. How much trash did Rashid collect rounded to the nearest tenth of a kilogram?

EXERCISE 2-6

Solve each subtraction problem as indicated.

1. What is the difference between 0.87 and 0.53? Draw a hundreds grid to solve.

2. What is the difference between 17 and 10.825? Use a vertical format to solve.

3. Ben has $43.75 in cash. He wants to buy a new wireless keyboard for his computer that costs 38.99. How much money will Ben have after his purchase?

4. On the first day of the month, Juanita has $2,385.77 in her checking account. She writes checks for $322.85 and $875 during the first week of the month. She makes a deposit of $912.45 on the 15th of the month. How much money does Juanita have in her checking account on the 15th of the month?

EXERCISE 2-7

Use what we've learned so far to solve these multiplication problems.

1. What is the product of 7 and 0.45? Use a vertical format to solve.

2. What is the product of 4.2 and 0.65? Use a vertical format to solve.

3. Jocelyn bought 2.5 pounds of cookies at a cost of $14.50 per pound. How much did Jocelyn pay for the cookies?

4. The fastest moving ant can walk 33.66 inches in a second. How many inches can this ant travel in 2.25 seconds?

EXERCISE 2-8

Solve each division problem as indicated.

1. What is 487.32 divided by 12? Use a vertical long division format to solve.

2. What 0.702 divided by 0.78? Use a vertical long division format to solve.

3. A group of 8 hikers bought lunch, snacks, and water for a hiking trip. They spent a total of $102.80 on these items. If the hikers share costs evenly, how much does each hiker owe? Use a vertical long division format to solve.

4. Marcia has a pink ribbon that is 6.5 meters long. She wants to cut the ribbon into pieces that are 0.25 meters long. How many pieces will she have? Use a vertical long division format to solve.

Fractions

MUST KNOW

- A fraction is a number that represents part of a whole or part of a set.

- The numerator of a fraction represents how many parts of a whole are being considered. The denominator represents how many equal parts the whole is divided into.

- A factor is a number that when multiplied by another produces a given number.

- A multiple of a whole number is the product of the number and any nonzero whole number.

- We perform basic operations with mixed numbers by first changing them so they have the same denominator.

In daily life, we have many opportunities to deal with fractions. We may buy a pizza divided into slices. Each slice is a fraction of the whole. We may look at five pairs of socks, two-fifths of which are blue. We may go shopping based on sales such as "$\frac{1}{3}$ Off" or "$\frac{1}{2}$ Off!" When we speak of a specific time, we may use phrases with fractions, such as "half past the hour" and a "quarter to three." Can you think of other examples when we use fractions in daily life?

Understanding Fractions

When we think of a fraction, we most often think of it as part of a whole. When we see a cake divided into pieces, we are looking at a whole divided into parts. Another meaning of *fraction* is "part of a group or set of objects." If we see a vase with five tulips, two red and four yellow, then two-sixths of the tulips are red.

In the example shown below, the pizza was originally divided into eight equal slices, one of which is now missing. So, we can say $\frac{1}{8}$ of the pizza is missing or $\frac{7}{8}$ of the pizza remains. Taken together, the two fractions equal the whole pizza, or 1.

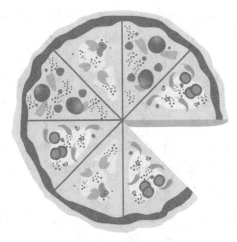

The rectangle below has been divided into 5 equal sections, two of which are shaded.

We can say $\frac{2}{5}$ of the rectangle is shaded or $\frac{3}{5}$ of the rectangle is unshaded. Once again, taken together $\left(\frac{2}{5} + \frac{3}{5}\right)$ equal the whole image (1).

All fractions can be represented as decimals. The simplest way to find the decimal value of a fraction is to divide the numerator by the denominator. Based on the pizza shown,

$$\frac{1}{7} = 0.\overline{142857} \text{ and } \frac{7}{8} = 0.875$$

Notice that in these examples, some decimal representations of fractions such as $\frac{7}{8}$ terminate, or come to an end. Others such as $\frac{1}{7}$ repeat indefinitely.

EXAMPLE

▶ Write the shaded portion of the diagram as a fraction and as a decimal.

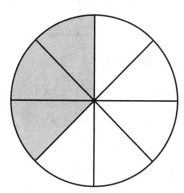

▶ The diagram shows $\frac{3}{8}$ of the circle shaded, and 3 divided by 8 equals 0.375: $\frac{3}{8} = 0.375$.

Multiples and Factors

When whole numbers other than zero are multiplied together, each number is a **factor** of the **product**. Thus, when we multiply 5 × 6 = 30, 5 and 6 are factors of the product 30. A **multiple** of a whole number is the product of the number and any nonzero whole number. 4, 8, 12, 16, 20, 24, etc. are examples of multiples of 4. Factors and multiples are essential to a good understanding of fractions.

Two or more numbers may share one or more multiples. The smallest of the common multiples shared by two or more numbers is called **the least common multiple (LCM)**. For example, consider the numbers 2 and 3. Their multiples are:

Multiples of 2: 2, 4, **6**, 8, 10, **12**, 14, 16, 18, 20, 22, 24, 26, **30**, ...

Multiples of 3: 3, **6**, 9, **12**, 15, 18, 21, 24, 27, **30**, ...

The numbers 6, 12, and 30 are multiples of 2 and 3. The number 6 is the least common multiple of the two numbers.

EXAMPLE

▶ What is the least common multiple of 6 and 15?

▶ Write multiples of each number and then see which is the smallest shared between the two groups.

Multiples of 6:   6, 12, 18, 24, **30**, 36

Multiples of 15:   15, **30**, 45, 60, 75

▶ So, the least common multiple of 6 and 15 is 30.

Once you find the least common multiple of two numbers, we can use this information to find a common denominator, therefore making it possible to add fractions with unlike denominators.

EXAMPLE

▶ What is the sum of $\frac{2}{6}$ and $\frac{5}{8}$?

▶ List the multiples of 6 and 8.

    Multiples of 6:   6, 12, 18, **24**, 30, 36, …

    Multiples of 8:   8, 16, **24**, 32, 40, …

▶ The least common multiple of 6 and 8 is 24.

▶ To add the two fractions, write them with a common denominator of 24.

$$\frac{2 \times 4}{6 \times 4} = \frac{8}{24} \qquad \frac{5 \times 3}{8 \times 3} = \frac{15}{24}$$

$$\frac{8}{24} + \frac{15}{24} = \frac{23}{24}$$

▶ The sum of $\frac{2}{6}$ and $\frac{5}{8}$ is $\frac{23}{24}$.

We also can use prime factorization to find the least common multiple of two or more whole numbers. Writing the **prime factorization** of a number involves writing the number as the product of its prime factors. A convenient way to do this is in the form a factor tree. For example, suppose we wanted to find the prime factorization of 30.

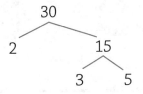

Write the original number.

Factor 30 as 2 × 15.

Factor 15 as 3 × 5.

So, the prime factorization of 30 is 2 × 3 × 5. We can easily check that this is correct using multiplication: 2 × 3 × 5 = 30.

> **EXAMPLE**
>
> ▶ What is the prime factorization of 45?
>
> ▶ Create a factor tree for 45. Notice that a factor, in this case 3, can occur more than once.
>
>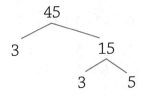
>
> ▶ The prime factorization of 45 is 3 × 3 × 5, or 3^2 × 5.

How can we use this information to find the least common multiple?

> **EXAMPLE**
>
> ▶ What is the least common multiple of 30 and 45?
>
> ▶ First, list the prime factors of each number and note the common factors.
>
> 30 = 2 × 3 × 5
> 45 = 3 × 3 × 5

- Multiply the highest power of each prime factor.

    $2 \times 3^2 \times 5 = 90$

- The least common multiple of 30 and 45 is 90.

A common factor is a whole number that is a factor of two or more nonzero numbers. **The greatest common factor (GCF)** is the greatest of the common factors of two or more numbers. We can solve some word problems by finding the GCF.

EXAMPLE

- Three groups of tourists buy entrance tickets to a theme park. The entrance tickets are the same price. The three groups pay $36, $54, and $81. What is the most a ticket costs?

- Find the GCF of the three amounts paid by listing all the factors of each amount spent.

    Factors of 36: 1, 2, 3, 4, 6, **9**, 12, 18, 36

    Factors of 54: 1, 2, 3, 6, **9**, 18, 27, 54

    Factors of 81: 1, 3, **9**, 27, 81

- The GCF of 36, 54, and 81 is 9. So, the most a ticket can cost is $9.

A number is **divisible** if the divisor divides the number without leaving a remainder. There are a number of divisibility rules that are helpful when you are factoring a number.

If the GFC of two numbers is 1, then neither is a factor or a multiple of the other.

Divisibility Rules	
2	If the ones digit is a 2, 4, 6, 8, or 0
3	If the sum of the digits is divisible by 3
5	If the ones digit is a 5 or 0
6	If the number is divisible by 2 and 3
9	If the sum of the digits is divisible by 9
10	If the ones digit is 0

IRL Divisibility rules have lots of practical applications in daily life. They are especially useful when we want to split a check evenly among friends!

EXAMPLE

▶ Is 359 divisible by 3? Which divisibility rule tells you the answer?

▶ The sum of the three digits is $3 + 5 + 9 = 17$.

▶ 17 is not divisible by 3.

▶ 359 is not divisible by 3 because the sum of its digits is not divisible by 3.

Here's another example.

EXAMPLE

▶ Is 738 divisible by 9? Which divisibility rule tells us the answer?

▶ The sum of the three digits is $7 + 3 + 8 = 18$.

▶ 18 is divisible by 9.

▶ 738 is divisible by 9 because the sum of its digits is divisible by 9.

Equivalent Fractions

Two different fractions that name the same number are called **equivalent fractions**. We can form equivalent fractions by multiplying or dividing the numerator and the denominator by the same number:

$$\frac{1 \times 2}{3 \times 2} = \frac{2}{6}, \text{ so } \frac{1}{3} \text{ and } \frac{2}{6} \text{ are equivalent fractions.}$$

$$\frac{24 \div 8}{32 \div 8} = \frac{3}{4}, \text{ so } \frac{24}{32} \text{ and } \frac{3}{4} \text{ are equivalent fractions.}$$

EXAMPLE

▶ What are two fractions that are equivalent to $\frac{3}{5}$?

$$\frac{3}{5} = \frac{3 \times 2}{5 \times 2} = \frac{6}{10}$$

$$\frac{3}{5} = \frac{3 \times 3}{5 \times 3} = \frac{9}{15}$$

▶ $\frac{6}{10}$ and $\frac{9}{15}$ are equivalent fractions to $\frac{3}{5}$.

Sometimes, we need to find a missing numerator or denominator to find equivalent fractions. To do this, we must figure out the relationship that exists between the two numerators or two denominators. For example,

$$\frac{5}{8} = \frac{20}{?} \qquad \text{Think:} \qquad 5 \times 4 = 20$$
$$8 \times 4 = ?$$

$8 \times 4 = 32$. Therefore, $\frac{5}{8} = \frac{20}{32}$.

> **EXAMPLE**
>
> ▶ Find the equivalent fractions of the following:
>
> $\dfrac{24}{56} = \dfrac{?}{7}$ Think: $56 \div 8 = 7$. If we also divide 24 by 8 we get 3:
>
> $\dfrac{24}{56} = \dfrac{3}{7}$
>
> ▶ $\dfrac{24}{56}$ and $\dfrac{3}{7}$ are, therefore, equivalent fractions.

A fraction is in its **simplest form** if its numerator and denominator have only 1 as a common factor; that is, no number other than 1 can divide both the numerator and the denominator.

> **EXAMPLE**
>
> ▶ In a survey of 100 teenagers, 64 percent prefer Brand X Jeans to Brand Y Jeans. Write the results of the survey as a fraction in its simplest form.
>
> ▶ Write the original fraction.
>
> $\dfrac{64}{100}$
>
> ▶ Divide the numerator and denominator by the GCF of both numbers.
>
> $\dfrac{64}{100} = \dfrac{64 \div 4}{100 \div 4} = \dfrac{16}{25}$
>
> ▶ In simplest form the fraction of students who prefer Brand X is $\dfrac{16}{25}$.

Putting fractions in order from least to greatest can be tricky when they all have different denominators. Here's how to go about solving this type of problem.

EXAMPLE

- Order the fractions $\frac{2}{3}$, $\frac{7}{10}$, and $\frac{3}{5}$ from least to greatest.

- Find the least common denominator (LCD) of the fractions. The LCM of 3, 10, and 5 is 30, so the LCD is 30.

- Write equivalent fractions using the LCD.

$$\frac{2}{3} = \frac{2 \times 10}{3 \times 10} = \frac{20}{30} \qquad \frac{7}{10} = \frac{7 \times 3}{10 \times 3} = \frac{21}{30} \qquad \frac{3}{5} = \frac{3 \times 6}{5 \times 6} = \frac{18}{30}$$

- The fractions in order from least to greatest are $\frac{3}{5}$, $\frac{2}{3}$, and $\frac{7}{10}$.

Adding and Subtracting Fractions

To add and subtract fractions with like denominators, we only need to follow three simple steps. First, add or subtract the numerators. Second, keep the denominator. Third, write the fraction in lowest terms.

EXAMPLE

- What is the sum of $\frac{5}{12}$ and $\frac{5}{12}$? Write the answer in its simplest form.

- Write the original problem.

$$\frac{5}{12} + \frac{5}{12} = ?$$

- Add the numerators. Keep the denominator the same.

$$\frac{5}{12} + \frac{5}{12} = \frac{10}{12}$$

- Simplify the answer to lowest terms.

$$\frac{10}{12} = \frac{10 \div 2}{12 \div 2} = \frac{5}{6}$$

> Reduced to its simplest form, the sum of $\frac{5}{12}$ and $\frac{5}{12}$ is $\frac{5}{6}$.

The same steps can be used when subtracting two fractions with the same denominator.

EXAMPLE

> What is the difference between $\frac{13}{15}$ and $\frac{8}{15}$? Write the answer in its simplest form.
>
> ▶ Write the original problem.
>
> $$\frac{13}{15} - \frac{8}{15} = ?$$
>
> ▶ Subtract the numerators. Keep the denominator the same.
>
> $$\frac{13}{15} - \frac{8}{15} = \frac{5}{15}$$
>
> ▶ Simplify the answer to lowest terms.
>
> $$\frac{5}{15} = \frac{5 \div 5}{15 \div 5} = \frac{1}{3}$$
>
> ▶ Reduced to its simplest form, the difference between $\frac{13}{15}$ and $\frac{8}{15}$ is $\frac{1}{3}$.

To add and subtract fractions with unlike denominators, we must first rename them as fractions with denominators that are the same. Then, we must simplify the fraction to its simplest form.

EXAMPLE

▶ What is the difference between $\frac{4}{5}$ and $\frac{1}{4}$? Write the answer in its simplest form.

▶ Write the original problem.

$$\frac{4}{5} - \frac{1}{4} = ?$$

▶ Use the LCM to write the fractions with a common denominator.

$$\frac{4 \times 4}{5 \times 4} = \frac{16}{20} \qquad \frac{1 \times 5}{4 \times 5} = \frac{5}{20}$$

▶ Subtract the fractions.

$$\frac{16}{20} - \frac{5}{20} = \frac{11}{20}$$

▶ Simplify the answer to lowest terms.

$\frac{11}{20}$ is in lowest terms.

▶ Reduced to its simplest form, the difference between $\frac{4}{5}$ and $\frac{1}{4}$ is $\frac{11}{20}$.

Adding and Subtracting Mixed Numbers

Some problems involve adding or subtracting mixed numbers. A **mixed number** consists of a whole number and a fractional part. In order to perform addition and subtraction, we must change the mixed number into an improper fraction. An **improper fraction** is any fraction in which the numerator is greater than the denominator.

EXAMPLE

▶ Jonathan ran $4\frac{1}{2}$ miles at a track meet. He then walked $1\frac{3}{5}$ miles to cool down. What is the total distance Jonathan traveled?

▶ Write the original problem.

$$4\frac{1}{2} + 1\frac{3}{5} = ?$$

▶ Find the LCD of the fractions.

$$\frac{1 \times 5}{2 \times 5} = \frac{5}{10} \qquad \frac{3 \times 2}{5 \times 2} = \frac{6}{10}$$

▶ Add the renamed fractions.

$$\frac{5}{10} + \frac{6}{10} = \frac{11}{10}$$

▶ Regroup the answer, if necessary.

$$\frac{11}{10} = 1\frac{1}{10}$$

▶ Add the whole numbers.

$$4 + 1 + 1 = 6$$

$$4\frac{1}{2} + 1\frac{3}{5} = 6\frac{1}{10}$$

▶ Jonathan traveled $6\frac{1}{10}$ miles.

Just like adding mixed numbers with fractional parts that have different denominators, sometimes we must rename to subtract mixed numbers.

EXAMPLE

▶ What is the difference between $3\frac{3}{5}$ and $2\frac{1}{3}$?

▶ Write the original problem.

$$3\frac{3}{5} - 2\frac{1}{3} = ?$$

▶ Find the LCD of the fractions.

$$\frac{3 \times 3}{5 \times 3} = \frac{9}{15} \qquad \frac{1 \times 5}{3 \times 5} = \frac{5}{15}$$

▶ Rename the fractions and subtract.

$$\frac{9}{15} - \frac{5}{15} = \frac{4}{15}$$

▶ Subtract the whole numbers.

$$3 - 2 = 1$$

▶ The difference between $3\frac{3}{5}$ and $2\frac{1}{3}$ is $1\frac{4}{15}$.

Multiplying and Dividing Fractions and Mixed Numbers

The steps for multiplying a fraction or a mixed number are quite simple:

- Write all the whole numbers and mixed numbers as fractions.
- Multiply the numerators of the fractions.
- Multiply the denominators of the fractions.
- Write the product in its lowest form.

EXAMPLE

▶ What is the product of $\frac{2}{5}$ and $\frac{1}{3}$? Write the product in its lowest terms.

▶ Write the original problem.

$$\frac{2}{5} \times \frac{1}{3} = ?$$

▶ Find the LCD of the fractions.

$$\frac{2 \times 3}{5 \times 3} = \frac{6}{15} \qquad \frac{1 \times 5}{3 \times 5} = \frac{5}{15}$$

▶ Multiply the numerators and then the denominators.

$$\frac{6}{15} \times \frac{5}{15} = \frac{30}{225}$$

▶ Reduce the product to it lowest terms.

$$\frac{30 \div 15}{225 \div 15} = \frac{2}{15}$$

▶ The product of $\frac{2}{5}$ and $\frac{1}{3}$ reduced to lowest terms is $\frac{2}{15}$.

Remember that we can write any whole number as a fraction by writing the number over a denominator of 1—for example, $4 = \frac{4}{1}$ and $17 = \frac{17}{1}$. Then, follow the rules by multiplying the numerators and the denominators and, finally, simplifying the product to lowest terms.

CHAPTER 3 Fractions

EXAMPLE

▶ Based on a class survey, $\frac{3}{5}$ of the 20 students in Mr. Yang's science class have one or more pets. As a fraction in lowest terms, how many students in Mr. Yang's science class have pets?

▶ Write the original problem.

$$\frac{3}{5} \times \frac{20}{1} = ?$$

▶ Multiply the numerators and then the denominators.

$$\frac{3}{5} \times \frac{20}{1} = \frac{60}{5}$$

▶ Reduce the product to lowest terms.

$$\frac{60 \div 5}{5 \div 5} = \frac{12}{1} = 12$$

▶ Therefore, 12 students in Mr. Yang's class have a pet.

When a problem asks us to multiply **mixed numbers**—numbers composed of a whole number and a fraction—we must first rewrite the numbers as improper fractions and then follow the same rules for multiplying fractions. Recall that an improper fraction is a fraction in which the numerator is greater than the denominator.

EXAMPLE

▶ What is the product of $4\frac{1}{3}$ and $3\frac{3}{4}$?

▶ Change the mixed numbers into improper fractions.

$$4\frac{1}{3} = \frac{13}{3} \qquad 3\frac{3}{4} = \frac{15}{4}$$

- Multiply the numerators and then the denominators.

$$\frac{13}{3} \times \frac{15}{4} = \frac{195}{12}$$

- Reduce the product to lowest terms.

$$\frac{195 \div 12}{12 \div 12} = 16\frac{3}{12} = 16\frac{1}{4}$$

- The product of $4\frac{1}{3}$ and $3\frac{4}{5}$ reduced to lowest terms is $16\frac{1}{4}$.

Once you have mastered the multiplication of fractions and mixed numbers, learning to divide them involves just one more concept or idea—the reciprocal. Recall that two numbers are reciprocals when their product equals 1. For example, $\frac{2}{5}$ and $\frac{5}{2}$ are reciprocals because $\frac{2}{5} \times \frac{5}{2} = \frac{10}{10} = 1$. To form the reciprocal of any fraction, all we have to do is switch the numerator and denominator. Thus, the reciprocal of $\frac{4}{9}$ is $\frac{9}{4}$ and the product of both is $\frac{36}{36} = 1$.

EXAMPLE

- What is the quotient of $\frac{4}{9}$ and $\frac{2}{3}$?

- Write the original problem.

$$\frac{4}{9} \div \frac{2}{3} = ?$$

- Rewrite as a multiplication problem, using the reciprocal of the divisor.

$$\frac{4}{9} \times \frac{3}{2} = ?$$

- Multiply the numerators and then the denominators.

$$\frac{4}{9} \times \frac{3}{2} = \frac{12}{18}$$

▶ Reduce the quotient to lowest terms.

$$\frac{12 \div 6}{18 \div 6} = \frac{2}{3}$$

▶ The quotient of $\frac{4}{9}$ and $\frac{2}{3}$ reduced to lowest terms is $\frac{2}{3}$.

Remember that to divide a fraction by a whole number, we simply rewrite the whole number as the numerator and place it over the denominator 1. Dividing mixed numbers, however, can be a little more complicated.

EXAMPLE

▶ At breakfast, Max eats $1\frac{1}{2}$ ounces of cereal. A full box of the cereal contains $16\frac{1}{2}$ ounces. How many $1\frac{1}{2}$ ounce servings are in a full box of cereal?

▶ Write the original problem.

$$16\frac{1}{2} \div 1\frac{1}{2} = ?$$

▶ Rewrite the mixed numbers as improper fractions.

$$\frac{33}{2} \div \frac{3}{2} = ?$$

▶ Rewrite the problem as a multiplication problem. Use the reciprocal of the second fraction. Then, multiply the numerators and the denominators.

$$\frac{33}{2} \times \frac{2}{3} = \frac{66}{6}$$

- Reduce the quotient to lowest terms.

$$\frac{66 \div 6}{6 \div 6} = \frac{11}{1}$$

- $16\frac{1}{2}$ divided by $1\frac{1}{2}$ is 11. There are, therefore, eleven $1\frac{1}{2}$ ounce servings of cereal in a full box.

Sometimes, it's a good idea to estimate the quotient when dividing fractions and mixed numbers. You can do this in the same way you estimate products when multiplying.

EXAMPLE

- What is a reasonable estimate of the quotient of $\frac{8}{9}$ and $\frac{1}{3}$?
- Rewrite as a multiplication problem, using the reciprocal of the divisor.

$$\frac{8}{9} \times \frac{3}{1} = ?$$

- Round to find approximate numbers. Note that $\frac{8}{9}$ is close to 1, and 3 equals 3.
- Estimate: There are approximately three $\frac{1}{3}$s in $\frac{8}{9}$.
- A reasonable estimate of the quotient of $\frac{8}{9}$ and $\frac{1}{3}$ is $\frac{1}{3}$.

Real-world problems may require us to figure out whether one amount fits a certain requirement.

CHAPTER 3 Fractions

EXAMPLE

▶ Ray and his mother are making decorative pillows as holiday gifts for eight family members. They have $5\frac{1}{2}$ yards of fabric. Each pillow takes about $\frac{2}{3}$ of a yard of fabric. Do Ray and his mother have enough fabric to make the eight pillows?

▶ Write the facts as a division problem.

$$5\frac{1}{2} \div \frac{2}{3} = ?$$

▶ Rewrite the mixed numbers as improper fractions.

$$\frac{11}{2} \div \frac{2}{3} = ?$$

▶ Rewrite the problem as a multiplication problem. Use the reciprocal of the second fraction. Then, multiply the numerators and the denominators.

$$\frac{11}{2} \times \frac{3}{2} = \frac{33}{4}$$

▶ Pull out whole number to get $8\frac{1}{4}$.

▶ The quotient of $5\frac{1}{2}$ and $\frac{2}{3}$ is $8\frac{1}{4}$. Ray and his mother, therefore, can make eight complete decorative pillows.

EXERCISES

EXERCISE 3-1

For each figure below, determine what fraction of the whole has been colored in.

1.

2.

EXERCISE 3-2

Find the least common multiple (LCM) of each pair of numbers.

1. 9 and 12

2. 14 and 21

EXERCISE 3-3

Write the prime factorization of each number.

1. 60

2. 176

EXERCISE 3-4

Find the greatest common factor (GCF) of each pair of numbers.

1. 12 and 15

2. 30 and 90

EXERCISE 3-5

Find the difference between the fractions in each question.

1. What is the difference between $\frac{1}{6}$ and $\frac{1}{8}$?

2. What is the difference between $\frac{3}{4}$ and $\frac{2}{7}$?

3. For a woodworking project, Fran nails together two pieces of wood. One piece is $\frac{2}{5}$ inch thick and the other is $\frac{1}{2}$ inch thick. In lowest terms, what is the combined thickness of the pieces of wood after being nailed?

4. On Saturday, $\frac{3}{5}$ inch of snow fell. On Sunday, $\frac{1}{3}$ inch of snow fell. How much more snow fell on Saturday than Sunday?

EXERCISE 3-6

Find the sum or difference of the mixed numbers in the following questions.

1. What is the sum of $5\frac{3}{8}$ and $3\frac{3}{5}$?

2. One piece of gold weighs $5\frac{1}{2}$ grams and another gold piece weighs $2\frac{1}{3}$ grams. What is the weight of the two pieces of gold?

3. What is the difference between $4\frac{1}{2}$ and $2\frac{2}{5}$?

4. One stone weighs $4\frac{4}{5}$ ounces and another weighs $2\frac{2}{3}$ ounces. How many more ounces is the heavier stone than the lighter one?

EXERCISE 3-7

Find the product or quotient of the given fractions.

1. What is the product of $\frac{3}{8}$ and $\frac{4}{5}$ in lowest terms?

2. What is the quotient of $\frac{2}{9}$ and $\frac{3}{4}$ in lowest terms?

3. John is running a race that is $\frac{3}{4}$ miles long. He is leading at the $\frac{2}{3}$ point of the race. At this point, how far has John run (in lowest terms)?

4. You are making 2 pounds of a party mix consisting of nuts and raisins. If you put $\frac{1}{5}$ pound of the party mix into each bag, how many bags can you fill?

EXERCISE 3-8

Find the product or quotient of the mixed numbers in each question.

1. What is the product of $5\frac{1}{3}$ and $2\frac{1}{4}$?

2. What is the quotient of $8\frac{1}{2}$ and $2\frac{3}{4}$?

3. A cookie recipe calls for $2\frac{2}{3}$ cups of sugar. If a baker is making $2\frac{1}{2}$ times the number of cookies, how much sugar is needed?

4. An adult cat that weighs 8 pounds needs about $\frac{4}{5}$ of a cup of dry food daily. If a bag of dry cat food contains $10\frac{1}{2}$ cups, about how many days will it feed the cat?

5. It takes $18\frac{3}{4}$ inches of ribbon to decorate a certain size gift box. If you have 100 inches of ribbon in all, how many gift boxes of this size can you decorate?

Flashcard App

Integers

MUST KNOW

⚡ An integer is a whole number that is positive, negative, or zero.

⚡ Positive integers are to the right of 0 and negative numbers are to the left of 0 on a number line.

⚡ On a number line, an integer is greater than a number on its left and less than a number on its right.

⚡ We can perform basic operations with integers by using a number line, absolute values, or rules for working with positive and negative numbers.

ver the years, you've almost certainly heard weather forecasters talk about record-breaking high or low temperatures. The highest recorded temperature in the United States was recorded in 1913 in California as 134°F. The lowest recorded temperature in the country was recorded in 1954 in Alaska as −80°F. Extreme temperatures like these are rare, but they do happen. The difference between these two temperatures is more than 200°F! These numbers are examples of integers.

Understanding Integers

An **integer** is a whole number that can be positive, negative, or zero. **Positive integers** are whole numbers greater than zero and have a positive (+) sign or no sign before them. **Negative integers** are whole numbers less than zero. Negative integers always have a negative sign (−) before them. Zero is considered neither a positive nor a negative integer. Two integers are considered **opposites** if they are the same distance from 0 on the number line but on opposite sides of 0. For example, 2 and −2 are opposites. The sum of two integers that are opposites is always 0.

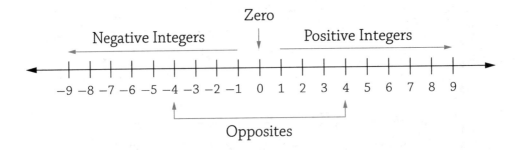

Integers are greater in value as you move from left to right along the number line, and they are lesser in value as you move from right to left. This makes comparing integers on a number line easy. Consider, for example, the numbers −3 and 2. Which is the greater number? When the points representing these numbers are marked on a number line, it's easy to see that 2 is greater than −3, because 2 is to the right of −3 and −3 is to the left of 2.

> **BTW**
> When reading aloud problems involving integers, remember that a number such as −5 is read as "negative 5" or, sometimes, "the opposite of 5."

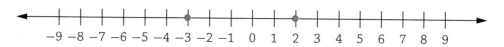

EXAMPLE

▸ Is 3 less than or greater than −4?

▸ Mark the numbers you are comparing on a number line. Check the position of each number.

▸ 3 is greater than −4, because 3 is to the right of −4.

Let's look at an example of how to compare two negative numbers on a number line. Remember that each number to the left of another number has a smaller value.

EXAMPLE

▸ Is −5 less than or greater than −7?

▸ Mark the numbers you are comparing on a number line. Check the position of each number.

▸ −5 is greater than −7, because −5 is to the right of −7.

The key to the next problem is to remember that opposites are the same number of units from 0, but in the opposite direction.

> **Which number is the opposite of 6?**
>
> ▸ Mark 6 on a number line.

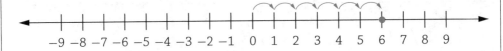

> ▸ Count the same number of units from zero in the opposite direction and mark that point.

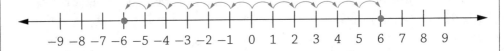

> ▸ Check the position of each number. So, the opposite of 6 is −6.

Adding Integers

There are several difference cases to consider when adding integers. You're well aware that adding two positive integers together results in a positive sum, for example: $5 + 9 = 14$. However, we do have to be careful not to confuse the addition sign and the sign for a positive integer. What the addition sentence above really says is: $(+5) + (+9) = (+14)$. What, however, happens when we add two negative integers?

For example, what is the sum of −4 and −3? We can use a number line to help us figure out the answer. Start out at 0 and move four spaces left to −4. Then, move three more spaces left.

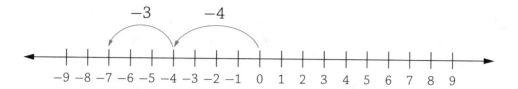

After marking the two numbers, we end up at −7, which is −3 and −4's sum.

Adding a positive and a negative number requires a little more thought. For example, what is the sum of −3 and 5? If we use a number line and follow the same steps as above, we'll find the correct answer. Start out at 0 and move three spaces left to −3. The next step is to add 5. Since the second addend is positive, we need to move five spaces to the right starting at −3: $-3 + 5 = 2$.

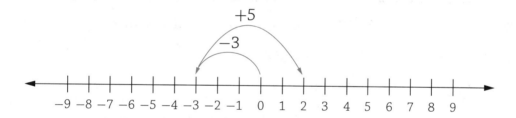

EXAMPLE

▶ What is the sum of −8 and 6? Use a number line to show how you found the answer.

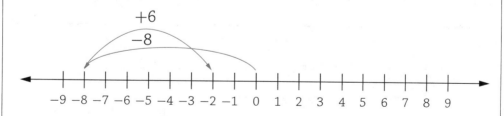

▶ The sum of −8 and 6 is −2.

> **BTW**
> When adding two integers with different signs, always identify the number with the greater absolute value first.

Using a number line is not the only way to add two integers. We can also use **absolute value**. The absolute value of a number is the distance between 0 and the number on a number line. So, -4 and 4 have the same absolute value: $|4|$. Notice that absolute value is written by placing the number between two vertical bars. Let's see how absolute value can be used to find the sum of $-8 + 5$:

$-8 = |8|$ and $5 = |5|$ First, find the absolute value of the addends.

$|8| - |5| = |3|$ Subtract the absolute values.

$|3| = -3$ Give the sum the sign of the greater addend.

EXAMPLE

▶ What is the sum of -12 and 9? Use absolute values to solve.

▶ Find the absolute value of the addends.

$-12 = |12|$ and $9 = |9|$

▶ Subtract the absolute values. Give the sum the sign of the greater addend in the original problem.

$|12| - |9| = |3|$

▶ Since $12 > 9$, use a negative sign: The sum of -12 and 9 is -3.

Let's look at an example where we add more than two integers with different signs.

EXAMPLE

▶ An underwater photographer dives to a depth of 50 feet. He then swims up toward the surface 15 feet. He dives another 10 feet down, and then dives further down by another 15 feet. What is the photographer's final location below water?

- Write the information as an equation involving addition.

$$x = -50 + 15 + (-10) + (-15)$$

- Find the sum of the negative numbers, and then find the sum of the positive numbers. Add the two sums. Give the final sum the sign of the greater addend.

$$x = -75 + 15$$
$$x = -60$$

- The diver's final location below water is -60 feet.

Subtracting Integers

No doubt, you're familiar with the idea that subtracting two whole numbers is the opposite of adding these numbers. The same is true of the addition and subtraction of integers. Simply stated, to subtract an integer we add its opposite. For example, what is the difference between 3 and 7?

$$3 - 7 = 3 + (-7)$$
$$= -4$$

The same principle is involved in finding the difference when the problem involves two negative numbers—for example, -5 and -8.

$$-5 - 8 = -5 + (-8)$$
$$= -13$$

EXAMPLE

- Solve: $40 - (-15)$.

- To subtract -15, add its opposite: $40 + (15) = 55$.

- The difference between 40 and -15 is 55.

Here's an example of a word problem that involves finding the difference of two negative integers.

> **EXAMPLE**
>
> ▸ The highest elevation in California is atop Mt. McKinley, which is about +14,500 feet above sea level. The lowest elevation is in Death Valley, which is about −280 feet below sea level. What is the difference in elevation between these two locations?
>
> ▸ Write the information as an equation involving subtraction.
>
> $x = +14{,}500 - (-280)$
>
> ▸ To subtract, add the opposite of −280.
>
> $x = +14{,}500 + 280$
> $x = 14{,}780$
>
> ▸ The difference in distance between these two locations is 14,780 feet.

Multiplying Integers

We multiply integers in the same way we multiply whole numbers, but with one extra step—we must keep track of the signs. The most foolproof method of doing this correctly is to multiply the absolute value of the numbers. Once we do that, we then deal with the signs by following two simple rules: The product of two positive integers or two negative integers is always positive, *and* the product of a positive and a negative integer is always negative. That's it! Let's look at a few examples of these rules in action.

EXAMPLE

▸ What is the product of −15 and −21?

▸ Write the absolute value of each number.

$$-15 = |15| \text{ and } -21 = |21|$$

▸ Multiply the absolute values.

$$|15| \times |21| = 315$$

▸ Check the sign of the product. The product of two negative integers is always positive, so −15 times −21 is 315.

Now, let's examine an example of multiplying two integers with different signs.

EXAMPLE

▸ What is the product of 14 and −6?

▸ Write the absolute value of each number.

$$14 = |14| \text{ and } -6 = |6|$$

▸ Multiply the absolute values.

$$|14| \times |6| = 84$$

▸ Check the sign of the product. The product of a positive and negative integer is always negative, so 14 times −6 is −84.

Let's look at a real-world example.

EXAMPLE

▶ Jada subscribes to an Internet service that costs $25 per month. The money is taken out of her checking account. After 12 months, how much will be subtracted from her checking account? Write the answer as a negative integer.

▶ Write the absolute value of each number.

$-25 = |25|$ and $12 = |12|$

▶ Multiply the absolute values.

$|25| \times |12| = |300|$

▶ Check the sign of the product. The product of a positive integer and a negative integer is always negative, so -25 times 12 is $-\$300$. Therefore, $\$300$ will be subtracted from Jada's checking account.

Dividing Integers

The rules for dividing integers are the same as those for multiplying. Therefore, the quotient of two integers with the same sign—either two positive signs or two negative signs—is always positive. Likewise, dividing two integers with different signs always results in a quotient that is negative.

EXAMPLE

▶ What is the quotient of -48 and -6?

▶ Write the absolute value of each number.

$-48 = |48|$ and $-6 = |6|$

▶ Divide the absolute values.

$|48| \div |6| = 8$

▶ Check the sign of the quotient. The quotient of two negative integers is always positive, so −48 divided by −6 is 8.

The following word problem presents an example of the division of integers with different signs:

EXAMPLE

▶ Randi's bank deducts a $15 per month fee if the balance on a checking account is less than $500. If $135 was deducted from Randi's account last year, in how many months was the balance less than the minimum of $500? Write the answer as a negative integer.

▶ Write the absolute value of each number.

$-15 = |15|$ and $-135 = |135|$

▶ Divide the absolute values.

$|135| \div |15| = 9$

▶ Check the sign of the quotient. The quotient of a negative integer and a positive integer is always negative, so −135 divided by −15 is −9. Randi's bank account had a balance of less than $500 for 9 months of last year.

EXERCISES

EXERCISE 4-1

Mark the numbers on the number line and compare.

1. Is 5 greater than or less than −2?

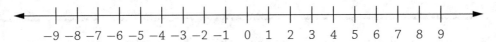

2. What is the opposite of −6?

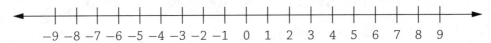

EXERCISE 4-2

Find the sum of the given integers.

1. What is the sum of −2 and −7? Draw a number line to help you answer the question.

2. What is the sum of −4 and +6? Draw a number line to help you answer the question.

3. What is the sum of 7 and −3? Use the absolute value of the numbers to find their sum.

4. On Saturday, Ron and Eileen hiked 5 miles in the morning and another 4 miles in the afternoon. On Sunday, they hiked 3 miles in the morning and 4 miles in the afternoon. How many miles did Ron and Eileen hike on Saturday and Sunday?

EXERCISE 4-3

Find the difference between the integers in each question.

1. $5 - (-11)$
2. $-8 - (+6)$
3. $28 - (+11)$

One more!

4. In the morning, Jeff climbs to the top of a mountain, which is 1,650 feet. Later that afternoon, he begins his descent and stops to eat after climbing down 550 feet. He rests again after climbing down another 445 feet. How high on the mountain is Jeff when he stops the last time?

EXERCISE 4-4

Find the product of the integers in each question.

1. -32×-17
2. -15×4
3. $-4 \times -9 \times -13$

Here's another one!

4. The temperature in a city dropped 3°F every hour from midnight until 5 A.M. If the temperature was 36°F at midnight, what was the temperature at 5 A.M.?

EXERCISE 4-5

Find the quotient of the integers in each question.

1. $-35 \div -7$

2. $63 \div -9$

3. $-156 \div -12$

Last question for the chapter!

4. The price of a share of stock decreased $12.50 over the course of a week. If the decrease was spread equally over five days of trading, how much did the price decrease each day?

Ratio and Proportion

MUST KNOW

⚡ A ratio uses division to make a comparison between two numbers.

⚡ We can compare ratios by writing them with the same denominator or by finding their decimal values.

⚡ A proportion is an equation stating that two ratios are equivalent.

⚡ A rate is a special type of ratio that compares two different types of quantities.

Ratios and proportions may sound like unfamiliar words, but you very likely have had experiences with these concepts in your daily life. If you're a baseball fan, you've dealt with ratios. How? Well, a player's batting average is the ratio of his hits to his times at bat. If you're told that a car is going 35 miles per hour, you've been given a ratio.

Understanding Ratios

A **ratio** uses division to make a comparison of two quantities. Each number in a ratio is called a term. Ratios can be written in three different ways. For example, if your school's science club has 7 boys and 8 girls as members, then the ratio of boys to girls in the science club can be expressed as 7 to 8, 7:8, or $\frac{7}{8}$. In all three cases, the ratio is read aloud as "7 to 8." Sometimes, a ratio can be simplified. For example, if the number of boys to girls in the science club is 6 to 8, you can simplify $\frac{6}{8}$ to $\frac{3}{4}$, or "3 to 4."

> ▶ There are 5 maple trees on the school property and 7 oak trees. What is the ratio of maple trees to oak trees?
>
> ▶ Think of the ratio as 5 maple trees to 7 oak trees, or $\frac{5 \text{ maple trees}}{7 \text{ oak trees}}$. Remember you can write a ratio in three different ways.
>
> ▶ The ratio of maple trees to oak trees is 5 to 7, 5:7, or $\frac{5}{7}$.

Let's look at another example.

> ▶ Your school's science lab has 24 microscopes and there are 36 students in your biology class. What is the ratio of microscopes to students?

CHAPTER 5 Ratio and Proportion **93**

- Think of the ratio as 24 microscopes to 36 students, or $\dfrac{24 \text{ microscopes}}{36 \text{ students}}$.
- Simplify the numerator and denominator.

$$\dfrac{24 \div 12}{36 \div 12} = \dfrac{2}{3}$$

- The ratio of microscopes to students is $\dfrac{2}{3}$.

Finding Equivalent Ratios

Equivalent ratios are ratios that represent the same comparison. For example, the ratio $\dfrac{3}{5}$ is equivalent, or equal, to $\dfrac{6}{10}, \dfrac{9}{15}, \dfrac{12}{20}$, and so on. To find equivalent ratios, we just need to multiply both terms by the same number.

EXAMPLE

- Raphael answered 4 out of 5 questions correctly on a daily math quiz. He answered 20 out of 25 questions correctly on a weekly math test. Is the ratio of correct answers to the total number of answers the same for both the quiz and the test?

- Write the two scores as ratios.

$$4 \text{ out of } 5 = \dfrac{4}{5}$$

$$20 \text{ out of } 25 = \dfrac{20}{25}$$

- Multiply the numerator and denominator by the same number.

$$\dfrac{4 \times 5}{5 \times 5} = \dfrac{20}{25}$$

▶ $\frac{4}{5}$ and $\frac{20}{25}$ are equivalent ratios. Therefore, the ratio of correct answers to all answers is the same on the quiz and the test.

Sometimes, we must use data from a table or other display to solve a problem involving equivalent ratios.

EXAMPLE

▶ A party punch contains 2 cups of orange juice for every 3 cups of pineapple juice. Complete the table below, which shows the ratios for different given quantities of punch.

Orange	2	4	6	?	10	?	14	?	18	?
Pineapple	3	?	?	12	?	18	?	24	?	30

▶ Use equivalent ratios to complete the table correctly. We can see that the each pair of numbers has a ratio of $\frac{2}{3}$. For example, then, we will fill in the box below the 4 with 6, the box below the 6 with 9, and so on.

▶ The ratios in the table are:

Orange	2	4	6	**8**	10	**12**	14	**16**	18	**20**
Pineapple	3	**6**	**9**	12	**15**	18	**21**	24	**27**	30

Comparing Ratios

There are two ways you can compare ratios to find which is greater or lesser. The first way is to write both ratios with the same denominator. Let's consider an example.

CHAPTER 5 Ratio and Proportion

> **EXAMPLE**
>
> ▶ Compare $\dfrac{3}{8}$ and $\dfrac{2}{5}$.
>
> ▶ Determine the value of each with the least common denominator:
>
> $$\dfrac{3\times 5}{8\times 5} = \dfrac{15}{40}$$
>
> $$\dfrac{2\times 8}{5\times 8} = \dfrac{16}{40}$$
>
> ▶ $\dfrac{2}{5} > \dfrac{3}{8}$ and $\dfrac{3}{8} < \dfrac{2}{5}$.

A simpler way to compare the value of two ratios is to find their decimal values by dividing the numerator by the denominator. It helps a lot to have a calculator handy when you use this method. Consider this question: Which is greater, $\dfrac{4}{7}$ or $\dfrac{5}{8}$?

$$\dfrac{4}{7} \approx 0.5714$$

$$\dfrac{5}{8} \approx 0.625$$

Since 0.625 is greater than 0.5714, $\dfrac{5}{8} > \dfrac{4}{7}$.

> **EXAMPLE**
>
> ▶ Jeff collects vintage DVDs. He has 15 comedy DVDs and 10 sci-fi DVDs. Jeff's best friend Raul has 13 sci-fi DVDs and 20 comedy DVDs. Whose collection has a greater ratio of sci-fi to comedy DVDs?
>
> ▶ Write the ratios for each collection as a fraction.
>
> Jeff's collection: $\dfrac{\text{sci-fi DVDs}}{\text{comedy DVDs}} = \dfrac{10}{15}$

Raul's collection: $\dfrac{\text{sci-fi DVDs}}{\text{comedy DVDs}} = \dfrac{13}{20}$

▸ Find the decimal value of each fraction.

$$\dfrac{10}{15} = \dfrac{2}{3} \approx 0.67$$

$$\dfrac{13}{20} = 0.65$$

▸ Since 0.67 > 0.65, Jeff's collection of DVDs has a greater ratio of sci-fi to comedy than Raul's.

Understanding Rates

A **rate** is a special type of ratio that compares two different types of quantities. When we hear that a car gets 30 miles per gallon, we are really being given a rate: $\dfrac{30 \text{ miles}}{1 \text{ gallon of gas}}$. The two quantities being compared are miles and gallons of gasoline. Other common examples of rates are 10 miles per hour, $20 per pound, and $15 per hour.

 IRL In expressions such as "30 miles per hour" and "60 minutes per hour," the word *per* means "for each."

▸ Franco is typing up a social studies report. He types 160 words in 5 minutes. What is the Franco's rate of words per minute in simplified form?

▸ Write the original rate given in the problem.

$$\dfrac{\text{words}}{\text{minutes}} = \dfrac{160}{5}$$

▸ Simplify the rate by using division.

$$\frac{\text{words}}{\text{minutes}} = \frac{160 \div 5}{5 \div 5} = \frac{32}{1}$$

▶ Franco types 32 words per minute.

When a rate compares how many units of one kind of quantity there are compared to one unit of another type of quantity, it is called a **unit rate**. Unit rates always have a denominator of 1. Unit rates can be very helpful when making comparisons.

▶ Danielle and Joachim both work at the same bakery. Danielle bakes 420 cookies in a 4-hour shift. During his 3-hour shift, Joachim bakes 330 cookies. Who baked more cookies per hour, Danielle or Joachim?

$$\text{Danielle: } \frac{\text{cookies}}{\text{hours}} = \frac{420 \div 4}{4 \div 4} = 105 \text{ cookies per hour}$$

$$\text{Joachim: } \frac{\text{cookies}}{\text{hours}} = \frac{330 \div 3}{3 \div 3} = 110 \text{ cookies per hour}$$

▶ Since 110 is greater than 105, Joachim bakes more cookies per hour than Danielle.

Solving Proportions

A **proportion** is an equation stating that two ratios are equivalent. As with ratios, the numbers in a proportion are called **terms**. The simplest way to determine if two ratios form a proportion is to determine if the cross products of its terms are equal. For example, you see that $\frac{3}{5}$ equals $\frac{12}{20}$, since 3 times 20 equals 60 and 5 times 12 is 60.

We can find an unknown number in a proportion by solving the proportion. Sometimes, this can easily be done using mental math. Take a look at the two numerators or denominators that appear in the problem to see if you can find a mathematical relationship. For example, when solving $\frac{123}{35} = \frac{x}{105}$, notice that 105 is three times 35. The unknown numerator is three times the known numerator; that is, 3 times 123 equals 369. So, x equals 369, or $\frac{123}{35}$ equals $\frac{369}{105}$.

Another way to find the value of the unknown term in a proportion is to use some basic algebra. Here's how you can go about using this method. What is the value of x in the proportion $\frac{5}{11} = \frac{x}{55}$?

First, write the original proportion.

$$\frac{5}{11} = \frac{x}{55}$$

Next, multiply each side by the denominator of the ratio with the unknown term x.

$$\frac{5}{11} \times \frac{55}{1} = \frac{x}{55} \times \frac{55}{1}$$

$$\frac{275}{11} = \frac{55x}{55}$$

Finally, solve for x by simplifying the fraction: $x = 25$.

EXAMPLE

▶ Solve the proportion: $\frac{22}{7} = \frac{x}{28}$. What is the value of x?

▶ Study the original proportion.

$$\frac{22}{7} = \frac{x}{28}$$

▶ Look for a relationship between the two known denominators. Notice that 7 times 4 equals 28.

CHAPTER 5 Ratio and Proportion 99

▶ Therefore, to keep the proportion, *x* is equivalent to 22 times 4 equals 88.

▶ Solve the proportion for the unknown value.

$$\frac{22}{7} = \frac{88}{28}$$

▶ Mutliply the cross products:

$22 \times 28 = 616$ and $7 \times 88 = 616$

▶ $\frac{22}{7} = \frac{88}{28}$ is a proportion because the ratios are equivalent.

Here's an example of a word problem that can be solved using proportions.

EXAMPLE

▶ Karen works for We Are Parties. She is blowing up balloons with helium for a party celebrting the birthday of her neighbor's child. If Karen can blow up 12 balloons in 15 minutes, how long will it take her to blow up 40 balloons?

▶ Write a proportion to represent the facts in the problem.

$$\frac{\text{balloons}}{\text{minutes}} \quad \frac{12}{15} = \frac{40}{x}$$

▶ Cross-multiply the terms of the proportion.

$12x = 15 \times 40$
$12x = 600$

BTW
When cross-multiplying, you are finding the product of the numerator of one term and the denominator of the other term.

▶ Solve the proportion for the unknown value.

$12x \div 12 = 600 \div 12$
$x = 50$

▶ It will take Karen 50 minutes to blow up 40 balloons.

EXERCISES

EXERCISE 5-1

For each problem, determine the ratio described.

1. In your school computer lab, there are 25 computers for every class of 40 students. What is the ratio of computers to students reduced to its simplest form?

2. An hour television show has 21 minutes of ads and 39 minutes of actual show. What is the ratio of show minutes to advertising minutes?

3. There are 46 students in the afterschool program. Of these, 24 are boys and 22 are girls. What is the ratio of girls to boys in the afternoon program? Write the ratio in its simplest form.

4. Jason has a rose garden. On a certain day, 16 yellow roses and 28 red roses are blooming. What is the ratio of red to yellow blooming roses?

EXERCISE 5-2

Let's apply what we've learned about ratios to some more problems.

1. By mid-season, your baseball team had won 8 out of 14 games. By the end of the season, it had won 12 out 21 games. Was the mid-season ratio of wins to games the same as the end-of-season ratio?

Fill in the blanks with $>$, $<$, or $=$.

2. $3:2$ ____ $54:36$

3. $9:12$ ____ $28:40$

4. $52:40$ ____ $36:27$

5. $4:17$ ____ $12:51$

EXERCISE 5-3

Solve the following problems by comparing the ratios.

1. In Mr. Jessup's 6th grade class, 23 out of 25 students are right-handed. In the school, there are 125 6th graders in all, and 110 are right-handed. Is the ratio of right-handed students in Mr. Jessup's class the same as it is for all 6th grade students?

2. Caleb and Malik collect basketball cards. Caleb has 23 New York Knicks cards and 17 Miami Heat cards. Malik has 24 New York Knicks cards and 18 Miami Heat cards. Whose collection of cards has the greater of ratio of New York Knicks to Miami Heat cards?

EXERCISE 5-4

Find the rate for each of these two problems.

1. The Judson family SUV consumed 13 gallons of gasoline when driving 325 miles in their new SUV. How many miles per gallon does the Judsons' SUV get?

2. Rachel and Juan both have part-time jobs at the Widget Factor. Rachel makes 210 widgets during her 3-hour shift, and Juan makes 260 widgets during his 4-hour shift. Who makes more widgets per hour, Rachel or Juan?

EXERCISE 5-5

For each problem, solve the proportion.

1. What is the value of x in the proportion $\frac{4}{12} = \frac{x}{39}$?

2. What is the value of k in the proportion $\frac{k}{15} = \frac{6}{45}$?

3. A cell-phone manufacturer's research shows that for every 300 new cell phones it makes, 2 are defective. At this rate, about how many newly manufactured cell phones out of 200,000 would be defective?

4. For a school raffle to raise money for library books, the ratio of raffle tickets you sold compared to your best friend was 12 : 5. If your friend sold 25 raffle tickets, how many raffle tickets did you sell?

Flashcard App

6 Percent

MUST KNOW

 A percent is a ratio that compares a number to 100.

 Every percent has a fraction and a decimal that is equivalent in value.

 The three cases of percent are: figuring out the percent of a number, determining what percent one number is of another, and finding a number when a percent of it is known.

 Problems involving sales tax, discounts, and interest are common real-world applications of percents.

We'd be hard pressed to experience a day in which percents are not mentioned at least once, if not several times. How often do you hear ads that boast about discounts such as "20% Off"? How about food items that pitch themselves as "99% Fat Free"? Businesses routinely brag about when sales or profits have increased 50%, 60%, even 100%. Nearly everyone has had some experience with the idea of the percent of interest. You pay interest in the form of a percent of the principle when you take out a car loan or a house mortgage. Better yet, you earn interest on financial transactions such as savings accounts and bonds.

A **percent** is a ratio that compares a number to 100. The word *percent* comes from Latin *per cento*, which means "out of 100." The symbol for percent is % and dates back to the Middle Ages.

A ratio can be represented in several different ways. You're already quite familiar with the idea of using a fraction to represent a ratio. A percent may be thought of as a fraction with a denominator of 100. Thus, $\frac{5}{25}$ may be represented by the equivalent fraction $\frac{20}{100}$, or $\frac{7}{10}$ as $\frac{70}{100}$. You can find the decimal equivalents of these fractions by dividing the numerators by the denominators: $\frac{20}{100}$ becomes 0.20 and $\frac{70}{100}$ becomes 0.70. Decimals can be written as percents by simply dropping the decimal point: 0.20 equals 20% and 0.70 equals 70%.

Solving Percent Problems Using Models

Since *percent* means "out of 100," it's convenient to use a model such as a hundreds grid to visually represent a percent. The three grids below show just three of the many ways we can do this:

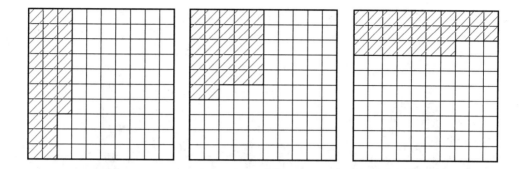

Percents may also be modeled using circle graphs. Let's estimate the percent each segment represents of the whole circle, below:

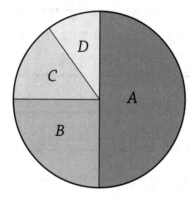

Section *A* is half, or 50%, of the whole circle. Section *B* is one-quarter of the circle, or 25%. Combined sections *C* and *D* represent another 25%. Since section *C* is larger than section *D*, a good estimate of section *C* is 15% and of section *D* is 10%.

Here are some other ways to represent fractions, decimals, and percents:

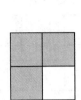
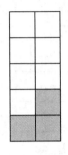
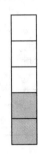

$\frac{3}{4} = 0.75 = 75\%$ $\frac{3}{10} = 0.3 = 30\%$ $\frac{2}{5} = 0.4 = 40\%$

EXAMPLE

▸ How do you write the shaded portion as a fraction, decimal, and percent?

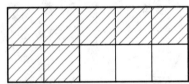

▸ Write a fraction that represents the diagram.

$$\frac{7}{10} = \frac{\text{shaded boxes}}{\text{all boxes}}$$

▸ Then, write a decimal that represents the fraction.

$$\frac{7}{10} = 0.70$$

▸ Finally, write a percent that represents the decimal.

$$\frac{7}{10} = 0.70 = 70\%$$

▸ The shaded portion of the diagram is written as $\frac{7}{10}$, 0.70, and 70%.

Sometimes we must use our understanding of geometric figures to find equivalencies among fractions, decimals, and percents.

> **EXAMPLE**
>
> ▶ How do you write the shaded portion as a fraction, decimal, and percent?
>
>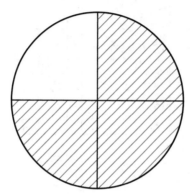
>
> ▶ Write a fraction that represents the diagram.
>
> $$\frac{3}{4} = \frac{\text{shaded sections}}{\text{all sections}}$$
>
> ▶ Then, write a decimal that represents the fraction.
>
> $$\frac{3}{4} = 0.75$$
>
> ▶ Finally, write a percent that represents the decimal.
>
> $$\frac{3}{4} = 0.75 = 75\%$$
>
> ▶ The shaded portion of the diagram is written as $\frac{3}{4}$, 0.75, and 75%.

Relating Fractions, Decimals, and Percents

You already know that you can write a ratio as a fraction or as a decimal. The grid below shows 35 of 100 squares shaded.

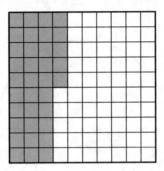

The ratio of these numbers can be represented by the fraction $\frac{35}{100}$ and by the decimal 0.35. Since a ratio compares a number to 100 and a percent means "out of 100," or "parts per hundred," it shouldn't come as a surprise that we can also represent a ratio as a percent. In this case, we can say 35% of the grid is shaded.

How do we represent the number of unshaded squares to the whole as a fraction, a decimal, and a percent?

$$\frac{65}{100} \; = \; 0.65 \; = \; 65\%$$

Fraction   Decimal   Percent

BTW

Grab a sheet of paper and make a three-column table showing common equivalencies among fractions, decimals, and percents. For example, write $\frac{1}{2}$ under the heading "Fraction," 0.5 under "Decimal," and 50% under "Percent." Then, do the same with other unit fractions such as $\frac{1}{3}$, $\frac{1}{4}$, and so on. Becoming comfortable with these equivalencies will speed up your calculations.

> **EXAMPLE**
>
> ▶ How do you write $\dfrac{31}{50}$ as a decimal?
>
> ▶ Multiply both numerator and denominator by 2.
>
> $$\dfrac{31}{50} \times \dfrac{2}{2} = \dfrac{62}{100} = 0.62$$
>
> ▶ $\dfrac{31}{50}$ is written as the decimal 0.62.

Here's an example that shows how to write a fraction as a percent.

> **EXAMPLE**
>
> ▶ How do you write $\dfrac{11}{25}$ as a percent?
>
> ▶ Multiply both numerator and denominator by 4.
>
> $$\dfrac{11}{25} \times \dfrac{4}{4} = \dfrac{44}{100} = 0.44 = 44\%$$
>
> ▶ $\dfrac{11}{25}$ is written as the percent 44%.

There are three common types, or **cases**, of percent. They involve finding:

- The percent of a number
- What percent one number is of another
- A number when the percent is known

Finding the Percent of a Number

The simplest percent problem to solve is to find the percent of a number. For example, suppose there are 40 books in a classroom library and 30% of them are nonfiction. How many books in the classroom library are nonfiction?

As you know, every percent can be written as a decimal or as a fraction, so there are two ways to solve the problem:

$$\begin{array}{c} 40 \\ \times 0.3 \\ \hline 12.0 \end{array} \quad \text{or} \quad 40 \times \frac{30}{100} = \frac{40}{1} \times \frac{30}{100} = \frac{1{,}200}{100} = 12$$

Either method is correct, but most students find the decimal method easier than the dealing with fractions. Use whichever method you are more comfortable with.

> **EXAMPLE**
>
> ▶ What is 15% of 120?
>
> ▶ Multiply 120 by 0.15.
>
> $$\begin{array}{r} 120 \\ \times\ 0.15 \\ \hline 600 \\ 1200 \\ \hline 18.00 \end{array}$$
>
> ▶ 15% of 120 is 18.

Here's an example of a word problem that involves finding the percent of a number.

> **EXAMPLE**
>
> ▶ Gillian bought a used car for $2,400. There is a 6% sales tax on the purchase. How much sales tax did Gillian pay?

CHAPTER 6 Percent   **111**

▸ Multiply $2,400 by 6%. 6% equals 0.06.

```
    2400
  × 0.06
  ──────
   14400
    0000
    0000
  ──────
  144.00
```

▸ Gillian paid $144.00 in sales tax.

BTW

A percent problem may seem more difficult than it actually is. For example, suppose you want to find 66% of 50. Rather than do the original calculation, simply flip the problem to finding 50% of 66. Since 50% is equivalent to $\frac{1}{2}$, all you have to do is think: Half of 66 is 33. So, 66% of 50 is 33!

Sometimes, the percent in a problem is greater than 100%.

EXAMPLE

▸ What is 125% of 80?

▸ Write 125% as 1.25, and multiply 80 by 1.25.

```
       80
   × 1.25
   ──────
      400
      160
       80
   ──────
   100.00
```

or $\frac{80}{1} \times \frac{125}{100} = \frac{10,000}{100} = \frac{100}{1} = 100$

▸ 125% of 80 is 100.

Finding What Percent One Number Is of Another Number

Another common type of percent problems asks us to find what percent one number is of another. For example, what percent of 150 is 42?

To solve this problem, we must write a percent equation:

$150 \times x\% = 42$ Write an equation.

$x\% = \dfrac{42}{150}$ Divide both sides by 150.

$x\% = 0.28$

$x = 28\%$ Change the decimal to a percent.

Let's work out a couple of examples together.

> **EXAMPLE**
>
> ▶ Eliza bought a electric motorized skateboard for $600. She paid $33 in sales tax. What percent was the tax rate?
>
> ▶ Let x equal the tax rate. Write an equation or proportion.
>
> $600 \times x\% = 33$
>
> $x\% = \dfrac{33}{600}$
>
> $x\% = 0.055$
>
> $x = 5.5\%$
>
> ▶ Eliza paid a sales tax rate of 5.5%.

The word problem below involves finding what percent one number is of another.

EXAMPLE

▶ Ari purchased a vintage six-CD set of his favorite country music singer. The set was originally priced at $90, but Ari bought it on sale for $72. What percent of the original price was the sale price?

▶ Let x equal the sales rate. Write an equation or proportion.

$$90 \times x\% = 72$$
$$x\% = \frac{72}{90}$$
$$x\% = 0.8$$
$$x = 80\%$$

▶ The sale price was 80% of the original price.

Finding a Number When the Percent Is Known

Another common percent problem involves finding a number when a percent of the number is known. Suppose a video game manufacturer knows that it sold about 43,500 copies of its newest game to teenagers last year. The company thinks sales to teens account for 80% of all copies of the game it sold. How many video games did the manufacturer sell in all?

To solve this problem, we can write an equation and solve. Think: 80% of what number is 43,500? Write 80% as 0.8:

$$x = 43{,}500 \div 0.8$$
$$x = 54{,}375$$

Therefore, the manufacturer sold a total of 54,375 copies of its new video game.

> **EXAMPLE**
>
> ▶ If 60% of a number is 1,050, what is the number?
>
> ▶ Write an equation and solve.
>
> $$0.6 \times x = 1050$$
> $$x = \frac{1050}{0.6}$$
> $$x = 1{,}750$$
>
> ▶ 1,050 is 60% of 1,750.

Here's example that shows how to find a number when the percent is known.

> **EXAMPLE**
>
> ▶ Justina left a 15% tip of $6.50 on a dinner bill at her favorite restaurant. About how much was Justina's dinner bill?
>
> ▶ Think of 15% as 0.15. Write an equation.
>
> $$0.15 \times x = 6.50$$
> $$0.15x = 6.50$$
> $$\frac{0.15x}{0.15} = \frac{6.50}{0.15}$$
> $$x = 43.33$$
>
> ▶ Justina's dinner bill was about $43.33.

Finding Percent of Increase or Decrease

The **percent of change** shows how much a quantity has increased or decreased when compared to the original amount. The **percent of increase** tells us the percent change when the new amount is greater than the original amount. For example, suppose a local clothing store sold $8,100 worth of clothing last month. This month the store sells $9,120 worth of clothing. What is the percent of increase in sales (to the nearest whole percent)?

To find the answer subtract last month's sales from this month's sales: $9,120 - $8,100 = $1,020$. Then, write a ratio:

$$\frac{\text{amount of increase}}{\text{original number}} = \frac{\$1,020 \div 30}{\$8,100 \div 30} = \frac{34}{270} \approx 0.13 \approx 13\%$$

Therefore, the percent of increase in sales is 13%.

> **EXAMPLE**
>
> ▶ An Internet service announced an increase in its monthly subscriber price from $25 to $32. What is the percent of increase for a monthly Internet subscription from the provider?
>
> ▶ Find the dollar amount of the increase by subtracting: $32 - 25 = 7$.
>
> ▶ Write a percent of change equation.
>
> $$\frac{\text{amount of increase}}{\text{original number}} = \frac{7}{25} = 0.28$$
>
> ▶ The percent of increase in the Internet monthly subscription fee is 28%.

The **percent of decrease** indicates the percent change when the new amount is less than the original amount. Let's look at an example.

> **EXAMPLE**
>
> ▶ In September, a one-way airfare from Chicago to Los Angeles was $250. The price was reduced in December to $175. What was the percent of decrease in this airline's fare for this trip?
>
> ▶ Find the dollar amount of the decrease by subtracting: $250 - 175 = 75$.

- Write a percent of change equation.

$$\frac{\text{amount of decrease}}{\text{original number}} = \frac{75}{250} = 0.3$$

- The percent of decrease in the airfare is 30%.

Finding Simple Interest

Interest is payment for the use of money. If we borrow money from a bank for a period of **time** (t), we pay an annual **rate** (r) of interest based on the **principal** (P), or the amount we borrowed. If we deposit money in a savings account at the bank, we will receive interest based on a rate and the amount of time we leave the money in the bank.

For example, suppose you deposit $1,500 in a savings account at 2.5% annual simple rate of interest. You leave the money in the account for two years. How much interest do you earn? The best way to find the simple interest on the account is to use the formula: Interest (I) = principal (P) × rate (r) × time (t):

$$I = \$1{,}500 \times 2.5\% \text{ per year} \times 2 \text{ years} = \$1{,}500 \times 0.025 \times 2$$
$$I = \$75$$

In two years, you will earn $75 in interest on a savings account of $1,500 at a simple annual interest rate of 2.5%.

EXAMPLE

- You borrow $10,000 to buy a car. The bank offers you a four-year loan at 8% simple interest. How much interest will you pay on the loan at the end of the four years?

- Use the formula for finding simple interest: $I = Prt$.
- Substitute values from the problem. Use a decimal to represent the annual interest rate.

$$I = \$10{,}000 \times 8\% \text{ per year} \times 4 \text{ years} = \$10{,}000 \times 0.08 \times 4$$
$$= \$3{,}200$$

- The interest paid on the car loan after 4 years is $3,200.

EXERCISES

EXERCISE 6-1

Answer each question using the given figure.

1. Write the shaded part of the circle as a fraction, a decimal, and a percent.

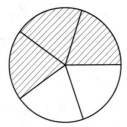

2. Write the unshaded part of the diagram as a fraction, a decimal, and a percent.

EXERCISE 6-2

Answer each question using the given figures.

1. What is a reasonable estimate of the percent of section *C* to the whole circle? Explain your reasoning.

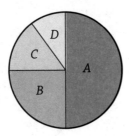

2. What is a reasonable estimate of the percent of the unshaded portion to the whole grid? Explain your reasoning.

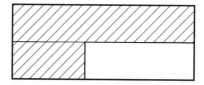

EXERCISE 6-3

For each problem, write the fraction, decimal, or percent indicated.

1. $\dfrac{19}{25}$ as a decimal

2. $\dfrac{13}{20}$ as a percent

3. 45% as a fraction

4. 55% as a decimal

EXERCISE 6-4

Solve the following percent problems.

1. What is 7% of 2,100?

2. Mr. Stone bought a painting for $500. Five years later he sold the painting for 140% of the purchase price. How much did Mr. Stone sell the painting for?

3. What percent of 375 is 75?

4. Sharona purchased a new sweater on sale. The original price of the sweater was $85, and Sharona bought it for $70. To the nearest whole number, what percent of the original price was the sale price?

5. If 55% of a number is 770, what is the number?

6. Franklin sold 105 cell phones this month. This represents an increase of 125% in sales from last month. How many cell phones did Franklin sell last month?

EXERCISE 6-5

Find the percent of increase or decrease.

1. During an end-of-year sale, the price of a pair of wireless airbuds was reduced from $400 to $250. What was the percent of decrease in the price of the airbuds?

2. A cable company raised its monthly service fee from $40 to $45. What is the percent of increase for the cable company's monthly service?

EXERCISE 6-6

Find the interest for each problem based on the information provided.

1. A deposit of $10,000 is placed in a college savings fund at a simple 2.0% annual rate of interest. How much interest will be earned on the account in four years?

2. You borrow $20,000 to renovate your kitchen. The bank has given you a five-year loan at a simple 8% annual rate of interest. How much interest will you have paid after three years?

Flashcard App

Equations and Inequalities

MUST KNOW

⚡ An equation tells us that two expressions are equal.

⚡ A function is a special mathematical relationship in which each input number in one set has exactly one output number in another set.

⚡ An inequality expresses a relationship between two numbers or quantities by using one of the inequality symbols: $<$, $\leq$, $>$, or $\geq$.

⚡ We can solve equations and inequalities using basic operations or graphing.

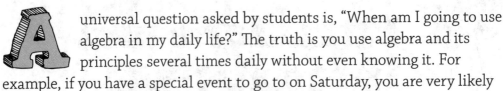A universal question asked by students is, "When am I going to use algebra in my daily life?" The truth is you use algebra and its principles several times daily without even knowing it. For example, if you have a special event to go to on Saturday, you are very likely to calculate how much time you need to get ready and how much the time it will take you to get there. In short, you would determine and combine two different values to come up with the answer. Another way to put it: $x = a + b$! Algebra and its methods come into play in virtually every profession—biology, chemistry, physics, computer sciences, even filmmaking, art, and architecture.

Defining Equation and Function

The most important word you will deal with in introductory algebra is *equation*. An **equation** is a mathematical sentence formed by equating, or placing an equal sign, between two expressions. The **solution** of an equation is the number that you substitute for a variable that makes the equation true.

For example, if you are carrying three books to school, their total weight can be expressed by the equation $x = a + b + c$. If textbook a weighs 2.4 pounds, textbook b weighs 3.6 pounds, and textbook c weighs 4.0 pounds, then their total weight, or the solution to the equation, is $x = 2.4 + 3.6 + 4.0$, or $x = 10.0$ pounds.

Whereas an equation has specific values that fit its meaning, a **function** is more general. A function pairs each input number in one set with exactly one output number in another set, for example, "each output number is 5 times the input number," which can be written as $y = 5x$. Suppose x equals the number of brownies and 5 equals the cost of $5 for each. Using the function, we can easily calculate the *cost of any number (y)* of brownies purchased—5 brownies cost $25, 8 brownies cost $40, and so on.

Reading and Writing Equations

In algebra, it's common to use a **variable** in the form of a letter such as a, n, x, or y to represent an unknown number. Expressions that contain at least one variable are called **variable expressions.** Some simple examples are the expressions $n + 2$, $a - 10$, $6 \times y$, and $x \div 5$. The meanings of these expressions are pretty obvious: the sum of a number plus 2, the difference between a number and 10, the product of a number times 6, and the quotient of a number divided by 5.

It's very helpful to become familiar with key words that indicate each operation that may be represented by a variable expression:

Operation	Key words
Addition	plus, added to, more than, sum of
Subtraction	minus, subtracted from, difference of
Multiplication	times, multiplied by, product of
Division	divided by, equal parts, quotient of

EXAMPLE
- Write a variable expression that represents 3 less than a number, n.
- The phrase "less than" indicates that this expression involves subtraction.
- The variable expression $n - 3$ represents "3 less than a number, n."

Here's another example of writing a variable expression.

EXAMPLE
- Write a variable expression that represents a number, x, divided by 2.
- The phrase "divided by" indicates that this expression involves division.
- The variable expression $x \div 2$ represents "a number, n, divided by 2."

Solving Equations

When we begin solving an equation, the very first thing we have to do is simplify it by combining like terms. **Like terms** are terms that have the same variable raised to the same power. For example, $4x - 4 - 3x = 0$ can be simplified as: $(4 - 3)x - 4 = 0 \rightarrow x - 4 = 0$.

Notice how the distributive property is used to combine the two terms that include x. Once we've combined the like terms, we're ready to take the next step in solving the equation. Remember that the goal is always to isolate the term x on one side of the equation and its value on the other side. In other words, we want an equation that says "x equals a certain value."

In the example above, to get x alone, we must get rid of -4. To do this, we can add $+4$ to both sides of the equation:

$$x - 4 = 0$$
$$x - 4 + 4 = 0 + 4$$
$$x = 4$$

Pretty simple, isn't it? There's just one more thing to do: check our answer by plugging it into the original equation to see if it gives us a true mathematical statement:

$$4x - 4 - 3x = 0$$
$$4(4) - 4 - 3(4) = 0$$
$$16 - 4 - 12 = 0$$
$$16 - 16 = 0$$
$$0 = 0$$

As long as we perform the same operation on both sides of an equation, it will always be in balance and equivalent to the original equation. It doesn't matter if we use addition, subtraction, multiplication, or division. Here's an example of an equation solved by using subtraction.

CHAPTER 7 Equations and Inequalities

EXAMPLE

▶ What is the value of x in the equation $x + 6 = 15$? Check the solution.

▶ Write the original equation:

$$x + 6 = 15$$

▶ Isolate x. Use the inverse operation. Undo the addition by subtracting 6 from both sides.

$$x + 6 - 6 = 15 - 6$$
$$x = 9$$

▶ Check the solution by plugging the value of x into the original equation.

$$x + 6 = 15$$
$$9 + 6 = 15$$
$$15 = 15$$

▶ The solution to the equation $x + 6 = 15$ is 9.

Recall that multiplication and division are inverse operations. We can use the fact that each operation undoes the other to solve equations.

EXAMPLE

▶ What is the value of t in the equation $7t = 35$? Check the solution.

▶ Write the original equation:

$$7t = 35$$

▶ Isolate t. Use the inverse operation. Undo the multiplication by dividing both sides of the equation by 7.

$$\frac{7t}{7} = \frac{35}{7}$$
$$t = 5$$

▶ Check the solution by plugging the value of *t* into the original equation.

$$7t = 35$$
$$7(5) = 35$$
$$35 = 35$$

▶ The solution to the equation $7t = 35$ is 5.

To solve some equations we will need to use two or more operations. Here's an example.

▶ It costs $35 per day plus $1.25 per mile to rent a car. If the car is rented for 2 days and is driven a total of 140 miles, how much will the car rental cost?

▶ Write an equation to represent the situation, where *C* equals the cost of the rental, *d* is the number of days, and *m* equals the number of miles driven.

$$C = 35d + 1.25m$$

▶ Plug the number of days and miles into the equation.

$$C = 35d + 1.25m$$
$$C = 35(2) + 1.25(140)$$

▶ Solve the equation to find *C*.

$$C = 35(2) + 1.25(140)$$
$$C = 70 + 175$$
$$C = \$245$$

▶ The total cost (*C*) of renting the car is $245.

CHAPTER 7 Equations and Inequalities 127

If an equation has a **coefficient**—the number by which a variable is multiplied—that is a fraction, we must use a reciprocal to find the solution. Here's an example that shows how to do this.

> **EXAMPLE**
>
> ▶ What is the value of x in the equation $\frac{5}{8}x - 3 = 2$? Check the solution.
>
> ▶ Write the original equation:
>
> $$\frac{5}{8}x - 3 = 2$$
>
> ▶ Isolate x. Use the inverse operation. Undo the subtraction by adding 3 to both sides.
>
> $$\frac{5}{8}x - 3 + 3 = 2 + 3$$
>
> $$\frac{5}{8}x = 5$$
>
> ▶ Use the reciprocal of $\frac{5}{8}$ to remove the coefficient of x.
>
> $$\left(\frac{8}{5}\right) \times \frac{5}{8}x = \left(\frac{8}{5}\right) \times 5$$
>
> $$x = \frac{40}{5}$$
>
> $$x = 8$$
>
> ▶ Check the solution by plugging the value of x into the original equation.
>
> $$\frac{5}{8}x - 3 = 2$$
>
> $$\frac{5}{8}(8) - 3 = 2$$

$$\frac{40}{8} - 3 = 2$$
$$5 - 3 = 2$$
$$2 = 2$$

▶ The solution to the equation $\frac{5}{8}x - 3 = 2$ is 8.

Graphing Equations

We've seen that an equation with one variable has only one solution. The equation $3x + 7 = 22$ is an example of this principle:

$$3x + 7 = 22$$
$$3x + 7 - 7 = 22 - 7$$
$$3x = 15$$
$$\frac{3x}{3} = \frac{15}{3}$$
$$x = 5$$

We can check the solution by plugging in 5 for x in the original equation:

$$3x + 7 = 22$$
$$3(5) + 7 = 22$$
$$15 + 7 = 22$$
$$22 = 22$$

How many solutions does a linear equation with two variables have? We can check this out by exploring an example.

CHAPTER 7 Equations and Inequalities

EXAMPLE

▶ What are three solutions to the equation $y = x + 2$?

▶ Choose three values for x and find the corresponding y-value of each. Make a table of your values.

x	y	(x, y)
−2	0	(−2, 0)
0	2	(0, 2)
2	4	(2, 4)

▶ Plot each point on a coordinate grid and draw a line through the points.

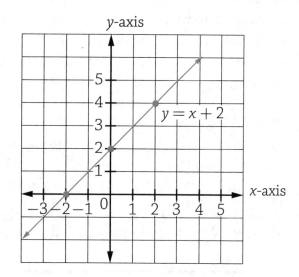

▶ Note that each point along the line is a solution to the equation. Based on the graph, you can determine that when x is 1, then y is 3.

▶ The linear equation $y = x + 2$ has an infinite number of solutions.

For linear equations in two variables, the input numbers for x are the **independent variables**; that is, they are simply given or chosen. The output numbers for y are **dependent variables** since their values are a function of the input values.

> **BTW**
> When choosing values for an input-output table, it's helpful to choose 0 and two numbers that are opposites on the number lines—for example, 5 and −5.

Monomials and Polynomials

A **monomial** is a one-term expression that can be a number, a variable, or the product of one or more variables. Examples of monomials include expressions such as 10, x, $5y$, $2x^2$, $8xy$, and $3x^2y$. Each monomial functions as single a term in an equation.

A **polynomial** is the sum of two or more monomials. Examples of polynomials include **binomials** with two terms such as $5x + 3$, $2x^2 + x^3y$, and $8xy + 5y$. **Trinomials** are polynomials with three terms—for example, $5x^3 + 2x^2 + 10x$ or $2x^2 + 4x + 8$. Sometimes, a polynomial is written as subtraction rather than addition; for example, $8xy - 5y$. You really should think of such expressions as the addition of a negative number: that is, $8xy - 5y$ really means $8xy + (-5y)$.

The **degree** of a polynomial refers to the value of the greatest exponent in any of its terms. For example, polynomial $5x^3 + 2x^2 + 10x$ has a "degree 3," whereas $2x^2 + 4x + 8$ has "degree 2." As a result, polynomials are usually written in standard order, meaning the term with the greatest exponent comes first, the next highest comes second, and so on.

EXAMPLE

▸ Write the polynomial $x^2 + 10x + 2x^4 + 4x^3$ in standard order. Identify the degree of the polynomial.

▸ Identify the exponent of each term and write the terms of the polynomial in descending order from the term with the greatest exponent to the one with the least exponent.

$2x^4$
$4x^3$
x^2
$10x$

▸ The standard form of this polynomial is $2x^4 + 4x^3 + x^2 + 10x$. Since the greatest exponent is 4, the polynomial has degree 4.

As we saw earlier, polynomials can be simplified by combining like terms.

EXAMPLE

▸ Simplify the polynomial below, write it in standard form, and identify its degree.

$$4k^2 + 10k + 3k^2 + (-4k^3) + (-5k)$$

▸ Combine like terms.

$$(4k^2 + 3k^2) + (10k - 5k) + (-4k^3)$$

$$7k^2 + 5k + (-4k^3)$$

▸ Write the terms of the polynomial in descending order from the term with the greatest exponent to the one with the least exponent.

$-4k^3$
$7k^2$
$5k$

▸ The simplified standard form of this polynomial is $-4k^3 + 7k^2 + 5k$. Since the greatest exponent is 3, the polynomial has degree 3.

Adding Polynomials

We can add and subtract polynomials using the same principles of simplification but using a vertical format.

EXAMPLE

▶ Add the polynomials $3x^2 - 8 + 6x$ and $2x - x^2 - 4$.

▶ Write each polynomial in standard form.

$3x^2 - 8 + 6x$ in standard form is $3x^2 + 6x - 8$.

$2x - x^2 - 4$ in standard form is $-x^2 + 2x - 4$.

▶ Set up the problem in vertical addition format.

$$\begin{array}{r} 3x^2 + 6x - 8 \\ +\ -x^2 + 2x - 4 \end{array}$$

▶ Add like terms.

$$\begin{array}{r} 3x^2 + 6x - 8 \\ +\ -x^2 + 2x - 4 \\ \hline 2x^2 + 8x - 12 \end{array}$$

▶ The sum of the two polynomials written in standard form is $2x^2 + 8x - 12$.

Multiplying Monomials

Just as we can add monomials and polynomials, we can also multiply them. Let's begin by looking at how to multiply two monomials. Recall that the commutative property of multiplication says that the order in which factors are multiplied does not change the product. In addition, the associative property of multiplication says that the grouping of factors also does not affect the product. Both these properties are used when multiplying two monomials. Here's an example of how it's done.

CHAPTER 7 Equations and Inequalities

EXAMPLE

▶ What is the product of the monomials $5x$ and $-4y$?

▶ Write the monomials as if they are factors in a multiplication problem.

$$(5x)(-4y)$$

▶ Use the commutative and associative properties of multiplication to regroup the coefficients and variables.

$$(5)(-4)(x)(y)$$

▶ Multiply like terms.

$$(5)(-4)(x)(y) = -20xy$$

▶ The product of the two monomials is $-20xy$.

When we multiply two monomials, we may need to use the laws of exponents. The product rule for exponents says that to multiply two powers having the same base, we must add the exponents, for example, $x^2 \times x^3 = x^{2+3} = x^5$.

EXAMPLE

▶ What is the product of the monomials $-3x$ and $-2x^2$?

▶ Write the monomials as if they are factors in a multiplication problem.

$$(-3x)(-2x^2)$$

▶ Use the commutative and associative properties of multiplication to regroup the coefficients and variables.

$$(-3)(-2)(x)(x^2)$$

▶ Multiply like terms.

$$(-3)(-2)(x)(x^2) = 6x^{1+2} = 6x^3$$

▶ The product of the two monomials $-3x$ and $-2x^2$ is $6x^3$.

The power rule for exponents says that, to find the power of a monomial that is itself a power, we must multiply the exponents, for example, $(3x^3)^2$ is $(3)^2 \times (x^3)^2 = 9x^6$.

> **EXAMPLE**
>
> ▶ What is the product of the monomials $2x^2y^2$ and $6x^2y$?
>
> ▶ Write the monomials as if they are factors in a multiplication problem.
>
> $(2x^2y^2)(6x^2y)$
>
> ▶ Use the commutative and associative properties of multiplication to regroup the coefficients and variables.
>
> $(2)(6)(x^2)(x^2)(y^2)(y)$
>
> ▶ Multiply like terms.
>
> $(2)(6)(x^2)(x^2)(y^2)(y) = 12x^4y^3$
>
> ▶ The product of the two monomials $2x^2y^2$ and $6x^2y$ is $12x^4y^3$.

Multiplying a Polynomial by a Monomial

When we multiply a polynomial by a monomial, we must use the distributive property as well as the laws of exponents. For example, by using the distributive property, the product of $4x(x + 3)$ becomes:

$4x(x + 3)$
$= 4x(x) + 4x(3)$
$= 4x^2 + 12x$

Here's a somewhat more complex example.

EXAMPLE

▶ What is the product of the monomial $-4a$ and the polynomial $2a^2 - 5a + 3$?

▶ Write the monomial and polynomial as if they are factors in a multiplication problem.

$$(-4a)(2a^2 - 5a + 3)$$

▶ Use the distributive property and the product rule for exponents to regroup the coefficients and variables.

$$(-4a)(2a^2) + (-4a)(-5a) + (-4a)(3)$$

▶ Multiply the grouped terms.

$$(-8a^3) + (20a^2) + (-12a) = -8a^3 + 20a^2 - 12a$$

▶ The product of the the monomial $-4a$ and the polynomial $2a^2 - 5a + 3$ is $-8a^3 + 20a^2 - 12a$.

Solving Inequalities

Recall that an **open sentence** in mathematics is a sentence that contains one or more variables that may be true or false depending on what values are substituted for the variables. An **inequality** expresses a relationship between two numbers or quantities by using one of the inequality symbols: $<, \leq, >$, or $\geq$. The meaning of these symbols is shown below.

Symbol	Meaning
$<$	is less than
$\leq$	is less than or equal to
$>$	is greater than
$\geq$	is greater than or equal to

Solving inequalities is very much like solving equations, but with one of the inequality symbols rather than an equal sign. In the next example, we can see how the inverse operations of addition and subtraction can be used to solve an inequality.

> ▶ What is the solution of the inequality $x + 4 \geq 5$?
>
> ▶ Write the inequality.
>
> $x + 4 \geq 5$
>
> ▶ Subtract 4 from both sides of the inequality.
>
> $x + 4 - 4 \geq 5 - 4$
>
> ▶ Solve the inequality.
>
> $x \geq 1$
>
> ▶ The solution to the inequality $x + 4 \geq 5$ is $x \geq 1$.

The inverse relationships of multiplication and division can also be used to solve an inequality. Here's an example of using division to solve an inequality.

> ▶ What is the solution of the inequality $8x - 2 \leq 46$?
>
> ▶ Write the original inequality.
>
> $8x - 2 \leq 46$
>
> ▶ Add 2 to both sides of the inequality.
>
> $8x - 2 + 2 \leq 46 + 2$

CHAPTER 7 Equations and Inequalities

▸ Divide both sides of the inequality by 8.

$$\frac{8x}{8} \leq \frac{48}{8}$$

$$x \leq 6$$

▸ The solution of the inequality $8x - 2 \leq 46$ is $x \leq 6$.

Another way to represent the solution to an inequality is by graphing it on a number line.

▸ What is the solution of the inequality $\frac{3}{4}y < 6$? Graph the solution on a number line.

▸ Write the original inequality.

$$\frac{3}{4}y < 6$$

▸ Multiply both sides of the inequality by the reciprocal $\frac{4}{3}$.

$$\frac{4}{3} \times \frac{3}{4}y < \frac{4}{3} \times 6$$

▸ Simplify and solve.

$$\frac{12y}{12} < \frac{24}{3}$$

$$y < 8$$

- Graph the solution on a number line. Notice that to show that 8 is not part of the solution, an open circle is used. (If 8 were part of the solution, then a closed circle would be used.)

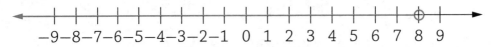

- The solution to the inequality $\frac{3}{4}y < 6$ is $y < 8$.

 IRL Graphing calculators and computer programs make it easy to graph equations and inequalities. However, remember that the accuracy of the graphs depends entirely on the accuracy of the input. Always keep in mind: Garbage in, garbage out!

CHAPTER 7 Equations and Inequalities

EXERCISES

EXERCISE 7-1

Write a variable expression that represents each of the following situations.

1. 4 more than a number n

2. a number, n, multiplied by 8

3. 2 less than a number n

4. a number, n, divided by 2

EXERCISE 7-2

Solve each of the following problems, and provide a check of the solution.

1. What is the value of x in the equation $x + 7 = 15$?

2. What is the value of y in the equation $6y = 54$?

3. A building that is 400 ft tall has a 75-ft long radio antenna atop its roof. The cable then runs from the top of the antenna through the building and 15 ft underground. How long is the cable?

4. What is the value of x in the equation $\frac{3}{5}x - 6 = 3$?

EXERCISE 7-3

Complete each step as described.

1. Fill in the input-output table below, to be used to solve the equation $y = x + 5$.

x	y	(x, y)
−2		
0		
2		

2. Draw a graph, and plot each point from the table in the previous question.

3. Draw a line on the graph to represent the equation.

4. Based on the graph, what is the value of y when $x = 1$?

EXERCISE 7-4

Identify the correct name of each item.

1. What is the expression $4y^3$ called?

2. What is the expression $2x^2 + 8x + 13$ called?

3. How is the polynomial $2x^2 + (-13) + 4x^4 + 5x$ written in standard form?

4. What is the degree of the polynomial in question 3?

EXERCISE 7-5

Solve each problems using the given operation.

1. What is the sum of the polynomials $5x^2 - 2 + 3x$ and $3x + 2x^2 + 5$ written in standard form?

2. What is the sum of the polynomials $4y^2 - y + 8$ and $2y^2 - 3y - 4$ written in standard form?

3. What is the product of the monomial $2x^2$ and the polynomial $5x^3 + 4$ written in standard form?

4. What is the product of the monomial $4k^3$ and the polynomial $k^3 + 11$ written in standard form?

EXERCISE 7-6

Solve each inequality.

1. $x + 8 \geq 12$

2. $\dfrac{2}{5}x \leq 6$

3. $8y \leq 56$. Graph the solution on a number line.

4. $5n - 10 > 10$. Graph the solution on a number line.

Flashcard App

Measurement and Geometry

MUST KNOW

⚡ In the U.S. measurement system, we use inches, feet, yards, and miles for length; ounces, pounds, and tons for weight; and fluid ounces, cups, pints, quarts, and gallons for capacity.

⚡ In both the U.S. and metric systems, the key units for measuring time are the second, minute, hour, day, week, and year.

⚡ In the metric system, we use millimeters, centimeters, meters, and kilometers for length; milligrams, grams, and kilograms for mass; and milliliters and liters for capacity.

⚡ Pi (π) is the ratio of the circumference of a circle to its diameter.

easurement has played a key role in people's daily lives since ancient times. Measurements of length were essential to marking out plots of land in ancient Egypt, India, and China. Weighing crops such as wheat and barley and measuring the capacity of vessels holding liquids such as water, wine, and oil helped promote trade and the rise of monetary systems. While each culture had its own terms, the essential importance of the measurement of length, weight, capacity, and time was everywhere.

Customary Units of Length

The measurement system used in the United States is called the U.S. customary system and is based on the old British system. You were first introduced to the common units of length in U.S. customary system in your early school years. By now you very likely have a good sense of what each measurement means. If you need some practice, find a ruler or a tape measure—or try walking a mile! Let's take a look at the essential customary measurements, their abbreviations, and their relationships.

Customary System: Length		
inch	in	$\frac{1}{12}$ foot = 1 inch
foot	ft	12 inches = 1 foot
yard	yd	3 feet = 1 yard
mile	mi	5,280 feet = 1 mile 1,760 yards = 1 mile

Let's take a look at an example, and start using our abbreviations!

CHAPTER 8 Measurement and Geometry

> **EXAMPLE**
>
> ▶ How many inches are there in $1\frac{1}{2}$ ft?
>
> ▶ There are 12 inches in a foot. Change $1\frac{1}{2}$ to 1.5 and then multiply: $12 \times 1.5 = 18$.
>
> ▶ There are 18 inches in $1\frac{1}{2}$ ft.

Here's an example of how to find the number of feet in a given number of yards.

> **EXAMPLE**
>
> ▶ How many feet are there in 3 yd?
>
> ▶ There are 3 ft in a yard. Multiply: $3 \times 3 = 9$.
>
> ▶ There are 9 ft in 3 yd.

Now let's look at an example of how to find the number of yards in a part of a mile.

> **EXAMPLE**
>
> ▶ How many yards are there in $\frac{1}{2}$ mi?
>
> ▶ A mile is 1760 yd. To find $\frac{1}{2}$ mi, divide by 2: $1760 \div 2 = 880$.
>
> ▶ There are 880 yd in $\frac{1}{2}$ mi.

If we need to find the number of inches in a certain number of feet, here's how to do it.

> **EXAMPLE**
>
> ▶ How many inches are there in 5 ft?
>
> ▶ A foot is 12 in and there are 5 ft. Multiply: $5 \times 12 = 60$.
>
> ▶ There are 60 inches in 5 ft.

BTW

The word *mile* comes from the Latin word *mille*, meaning "one thousand." Mille referred to 1,000 double paces; that is, one step taken by each foot as a person walks. This distance is about 5 ft. Therefore, a mile was a distance of about 5,000 feet.

Customary Units of Weight

The best way to find out your weight is to step on a scale. What you see there will give you the answer—that is, depending on where you are. If you weigh 100 pounds standing on Earth, you will weigh only $16\frac{1}{2}$ lb on the Moon!

The reason for the difference is that the moon's gravity is much less than Earth's.

There are three basic units of weight in the U.S. customary system—the ounce, the pound, and the ton. The abbreviations for and the relationships among these measures are shown in the following table.

\multicolumn{3}{c}{Customary System: Weight}		
ounce	oz	$\frac{1}{16}$ pound = 1 ounce
pound	lb	16 ounces = 1 pound
ton	T	2,000 pounds = 1 ton

CHAPTER 8   Measurement and Geometry

 IRL   The word *ton* has somewhat different meanings depending on the country you are working in and the system of measurement you are using. In the United States, a ton is 2,000 lb, but in Great Britain a ton weighs 2,240 lb. Furthermore, in the metric system, a ton weighs 1,000 kilograms (kg), which is a few pounds more than the U.S. ton. When working problems in your math class, always use the U.S. meaning unless you're told otherwise.

EXAMPLE

▶ How many ounces are there in 2 lb?

▶ A pound is 16 ounces and there are 2 lb. Multiply: $16 \times 2 = 32$.

▶ There are 32 oz in 2 lb.

The next example shows how to find the number of pounds that make up a given number of ounces.

EXAMPLE

▶ How many pounds do 56 oz make?

▶ A pound is 16 ounces and there are 56 oz. Divide: $56 \div 16 = 3.5$.

▶ There are 3.5, or $3\frac{1}{2}$, lb in 56 oz.

Now let's look at an example of how to find the number of tons that make up a given number of pounds.

EXAMPLE

▶ How many tons does a 13,000-lb truck weigh?

▶ A ton is 2,000 lb and there are 13,000 lb. Divide: $13,000 \div 2,000 = 6.5$.

▶ There are 6.5, or $6\frac{1}{2}$, T in 13,000 lb.

Let's take a look at one more example.

> **EXAMPLE**
>
> ▸ How many pounds does a 1.5-T car weigh?
>
> ▸ A ton is 2,000 lb and there are 1.5 T. Multiply: $1.5 \times 2{,}000 = 3{,}000$.
>
> ▸ There are 3,000 lb in 1.5 T.

Customary Units of Capacity

In English, the word *capacity* has two different meanings. When a manufacturer says that a car factory has the capacity of making 3,000 cars a day, it means that the factory can produce 3,000 cars every 24 hours. If the car maker says a particular car's gasoline tank has a capacity of 16 gallons, it means the tank can hold up to 16 gallons when it's full. In terms of measurement, capacity refers to the second meaning—how much can a container hold.

The basic units of capacity in the customary system are the fluid (or liquid) ounce, the cup, the pint, the quart, and the gallon. The following table shows the abbreviations for and the relationships among these measures.

Customary System: Capacity		
fluid ounce	fl oz	$\frac{1}{8}$ cup = 1 fluid ounce
cup	c	8 fluid ounces = 1 cup
pint	pt	2 cups = 1 pint
quart	qt	2 pints = 1 quart
gallon	gal	4 quarts = 1 gallon

CHAPTER 8 Measurement and Geometry

> **EXAMPLE**
>
> ▸ How many fluid ounces are there in 3 c?
>
> ▸ A cup holds 8 fl oz and there are 3 c. Multiply: $8 \times 3 = 24$.
>
> ▸ There are 24 fl oz in 3 c.

The next example shows how to find the number of cups in a certain number of pints.

> **EXAMPLE**
>
> ▸ How many cups are there in 3 pt?
>
> ▸ A pint holds 2 c and there are 3 pt. Multiply: $2 \times 3 = 6$.
>
> ▸ There are 6 c in 3 pt.

To find the number of pints in a given number of quarts, study the next example.

> **EXAMPLE**
>
> ▸ How many pints are there in 4 qt?
>
> ▸ A quart holds 2 pt and there are 4 qt. Multiply: $2 \times 4 = 8$.
>
> ▸ There are 8 pt in 4 qt.

Now let's look at an example of how to find the number of quarts in a given number of gallons.

> **EXAMPLE**
>
> ▶ How many quarts are there in 4 gal?
>
> ▶ A gallon holds 4 qt and there are 4 gal. Multiply: $4 \times 4 = 16$.
>
> ▶ There are 16 qt in 4 gal.

Understanding Units of Time

The most important thing to remember about understanding customary units of time is that it *isn't* the same thing as knowing how use the hands of a clock to read time or to read a digital time device such as a cell phone. The units that make up the time measurement system are the same in the U.S. customary system and the metric system. Since humans mark off a wide range of times, from seconds to centuries, there are a lot of measures that relate to each other, as the following table shows.

Customary and Metric Systems: Time		
second	s	$\frac{1}{60}$ minute = 1 second
minute	min	60 seconds = 1 minute
hour	hr	60 minutes = 1 hour
day	*	24 hours = 1 day
week	wk	7 days = 1 week
month	mo	28–31 days, or approx. 4 weeks = 1 month
year	yr	12 months, or 52 weeks, or 365 days = 1 year
decade	*	10 years = 1 decade
century	*	100 years, or 10 decades = 1 century

*For the most part, we don't tend to abbreviate "day," "decade," or "century."

> **EXAMPLE**
>
> ▶ How many seconds are there in 4 min?
>
> ▶ A minute is 60 s and there are 4 min. Multiply: $60 \times 4 = 240$.
>
> ▶ There are 240 s in 4 min.

Here's an example of how to find the number of minutes in a given number of hours.

> **EXAMPLE**
>
> ▶ How many minutes are there in $1\frac{1}{2}$ hr?
>
> ▶ An hour is 60 min and there are $1\frac{1}{2}$ hr. Change $1\frac{1}{2}$ to 1.5. Multiply: $60 \times 1.5 = 90$.
>
> ▶ There are 90 min in $1\frac{1}{2}$ hr.

If we need to find the number of hours in a certain number of days, here's how to do it.

> **EXAMPLE**
>
> ▶ How many hours are there in 2 days?
>
> ▶ A day is 24 hr and there are 2 days. Multiply: $24 \times 2 = 48$.
>
> ▶ There are 48 hr in 2 days.

The following example shows how to find the number of years in a certain number of months.

> **EXAMPLE**
>
> ▶ How many years are there in 36 months?
>
> ▶ A year is 12 months and there are 36 months. Divide: $36 \div 12 = 3$.
>
> ▶ There are 3 years in 36 months.

To find the number of years in a given number of decades, study the following example.

> **EXAMPLE**
>
> ▶ How many years are there in 5 decades?
>
> ▶ A decade is 10 years and there are 5 decades. Multiply: $5 \times 10 = 50$.
>
> ▶ There are 50 years in 5 decades.

The next example shows how to find the number of years in a given number of centuries.

> **EXAMPLE**
>
> ▶ How many years are there in $1\frac{1}{2}$ centuries?
>
> ▶ A century is 100 years and there are $1\frac{1}{2}$ centuries. Change $1\frac{1}{2}$ to 1.5. Multiply: $100 \times 1.5 = 150$.
>
> ▶ There are 150 years in $1\frac{1}{2}$ centuries.

CHAPTER 8 Measurement and Geometry

Metric Units of Length

The basic unit of length in the metric system is the meter. You know that in the U.S. system a yard is equal to 3 feet. A good way to visualize a meter is to think of it as slightly longer than a yard. More exactly, a yard is 36 inches long, and a meter is a little more than 39 inches.

All metric terms for length are formed from the word *meter* and a prefix. It's very helpful when working in the metric system to remember the meanings of these prefixes: *deci-* means "one-tenth," *centi-* means "one-hundredth," *milli-* means "one-thousandth," and *kilo-* means "one thousand."

Metric System: Length		
meter	m	100 centimeters or 1,000 millimeters = 1 meter
decimeter	dm	100 millimeters = 1 decimeter; 10 decimeters = 1 meter
centimeter	cm	10 millimeters = 1 centimeter; 100 centimeters = 1 meter
millimeter	mm	1,000 millimeters = 1 meter
kilometer	km	1,000 meters = 1 kilometer

EXAMPLE
▸ How many centimeters are there in 5 meters?
▸ A meter is 100 cm and there are 5 m. Multiply: 100 × 5 = 500.
▸ There are 500 cm in 5 m.

The next example shows how to find the number of decimeters in a given number of meters.

> **EXAMPLE**
>
> ▸ How many decimeters are there in 2.5 meters?
>
> ▸ A meter is 10 dm and there are 2.5 m. Multiply: $10 \times 2.5 = 25$.
>
> ▸ There are 25 dm in 2.5 m.

Now let's look at an example of how to find the number of kilometers when given a certain number of meters.

> **EXAMPLE**
>
> ▸ How many kilometers are there in 5,500 meters?
>
> ▸ A kilometer is 1,000 m and there are 5,500 m. Divide: $5,500 \div 1,000 = 5.5$.
>
> ▸ There are 5.5 km in 5,500 m.

The next example shows how to find the number of millimeters in a given number of meters.

> **EXAMPLE**
>
> ▸ How many millimeters are there in 3.6 meters?
>
> ▸ A meter is 1,000 mm and there are 3.6 m. Multiply: $1,000 \times 3.6 = 3,600$.
>
> ▸ There are 3,600 mm in 3.6 m.

To find the number of meters in a given number of kilometers, study the following example.

EXAMPLE

▸ How many meters are there in 2.9 kilometers?

▸ A kilometer is 1,000 m and there are 2.9 km. Multiply: $1{,}000 \times 2.9 = 2{,}900$.

▸ There are 2,900 m in 2.9 km.

BTW

To change kilometers to meters, simply move the decimal point three places to the right: 1.5 km becomes 1,500 m. Think of whole number measurements such as 2 km as 2.0 km.

Metric Units of Mass

You already know that how much you weigh in the U.S. system depends on where you are standing, because weight is determined by the force of gravity. This problem doesn't come up in the metric system because the metric system doesn't measure weight but, instead, measures **mass**, the quantity of matter in an object. (Despite this technical distinction, in everyday language *mass* and *weight* are interchangeable.)

All metric measurements of mass are based on the gram. As with metric length, all the terms used in metric mass are formed from the word *gram* and a prefix. The three key words to remember are *gram*, *milligram*, and *kilogram*. The following table gives their abbreviations and shows how these terms are related to each other.

Metric System: Mass		
gram	g	1 gram = $\frac{1}{1{,}000}$ of a kilogram
milligram	mg	$\frac{1}{1{,}000}$ gram = 1 milligram
kilogram	kg	1,000 grams = 1 kilogram; 1,000,000 milligrams = 1 kilogram

A good way to get a sense of each metric unit of mass is to make comparisons with everyday objects. For example, a paperclip has a mass of about 1 gram, a quart of milk has a mass of about 1 kilogram, and a small feather has a mass of a few milligrams.

> **EXAMPLE**
>
> ▶ Which metric unit would we use to measure the mass of a melon?
>
> ▶ We would use grams or kilograms to measure the mass of a melon.

Now, think about the measure of something with so little mass we likely wouldn't feel it in our hands.

> **EXAMPLE**
>
> ▶ Which metric unit would we use to measure the mass of a snowflake?
>
> ▶ We would use milligrams to measure the mass of a snowflake.

What about something so light in our hands, we know it's there but we can barely feel it?

> **EXAMPLE**
>
> ▶ Which metric unit would we use to measure the mass of a $10 bill?
>
> ▶ We would use grams to measure the mass of a $10 bill. In fact, all U.S. bills weigh one gram!

Now, try thinking about something very light but definitely in our hands, such as a coin.

> **EXAMPLE**
> ▶ If a nickel has a mass of 5 grams, what is its mass in milligrams?
> ▶ A gram is 1,000 mg and there are 5 g. Multiply: $1,000 \times 5 = 5,000$.
> ▶ There are 5,000 mg in 5 g.

How would we go about finding how many items we need to reach a certain mass?

> **EXAMPLE**
> ▶ If an apple has a mass of about 100 g, how many apples would have a mass of 1 kg?
> ▶ A kilogram is 1,000 g and an apple has a mass of 100 g. Divide: $1,000 \div 100 = 10$.
> ▶ 10 apples would have a mass of 1 kg.

BTW

In all, there are 20 different prefixes used in metric measures. For example, the prefix mega-, as in megaton, means an explosion equal to "a million tons" of TNT. The prefix giga-, as in gigabyte, means "a billion bytes" of computer memory. For very, very small measurements, micro- means "one-millionth" and nano- means "one-billionth." Can you figure out the meanings of microsecond and nanometer?

Metric Units of Capacity

Remember that in the U.S. system *capacity* refers to how much liquid a container can hold. Capacity means exactly the same in the metric system.

In everyday use, we really need to know the meaning of just two terms: *liter* and *milliliter*. The table shows how these two measures of capacity are related.

Metric System: Capacity		
liter	L	1 liter = 1,000 milliliters
milliliter	ml	$\frac{1}{1,000}$ liter = 1 milliliter

You're probably familiar with the capacity of a liter since many products, such as milk and water, are commonly sold in 1-liter bottles. The next time you hold one of these containers, check the label to see if it is a liter or, perhaps, a half-liter. Milliliters are much, much smaller units. A typical raindrop is about $\frac{1}{10}$ to $\frac{1}{20}$ of 1 milliliter (ml), so it takes about 10 to 20 raindrops to make a milliliter.

EXAMPLE

▶ Which metric unit would we use to measure the capacity of a family-size container of orange juice?

▶ We would use liters to measure the capacity of a family-size container of orange juice.

Did you ever hear a water drop coming from a faucet? One drop doesn't have much mass, does it? Now, take a look at the next example.

EXAMPLE

▶ Which metric unit would we use to measure the capacity of a drop of food flavoring?

▶ We would use milliliters to measure the capacity of a drop of food flavoring.

The next example shows how to relate two different measures of capacity.

> **EXAMPLE**
>
> ▶ If a glass has a capacity to hold 250 ml of water, how many glasses would hold a liter of water?
>
> ▶ A liter holds 1,000 ml and each glass holds 250 ml. Divide: 1,000 ÷ 250 = 4.
>
> ▶ It would take 4 glasses to hold 1 L of water.

Here's another example of how to relate two measures of capacity.

> **EXAMPLE**
>
> ▶ If a bucket of water has a capacity of 10 L, how many liters would it take to fill a sink that can hold 5 buckets of water?
>
> ▶ The bucket holds 10 liters and the sink can hold 5 buckets. Multiply: 10 × 5 = 50.
>
> ▶ The sink has a capacity of 50 L when it's full.

Now, let's apply what we've learned about measurement to some basic geometry.

Finding the Perimeter of a Polygon

As you'll recall, a polygon is a closed plane figure that is formed by joining three or more line segments called sides. Each side of a polygon meets exactly two other line segments. Polygons come in an infinite variety of shapes and sizes—triangles, quadrilaterals, and pentagons, just to name a few.

The **perimeter** of a polygon is the distance around the figure; that is, the sum of the measures of the lengths of all its sides. Finding the perimeter of a polygon is as simple as a geometry problem gets. All we have to do is add the lengths of all sides. When you write the sum of these measures, always be sure to use the unit of measure given in the problem.

EXAMPLE

▶ Find the perimeter of the rectangle shown.

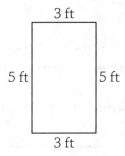

▶ Add the lengths of the four sides of the rectangle: $3 + 5 + 3 + 5 = 16$.

▶ The perimeter of the rectangle is 16 ft.

Now, let's try finding the perimeter of a triangle.

EXAMPLE

▶ Find the perimeter of the given triangle.

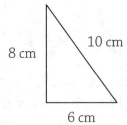

▶ Add the lengths of the three sides of the triangle: $6 + 8 + 10 = 24$.

▶ The perimeter of the triangle is 24 cm.

Some plane figures are irregular, but we find the perimeter in the same way as for regular polygons.

EXAMPLE

▶ Find the perimeter of the polygon below.

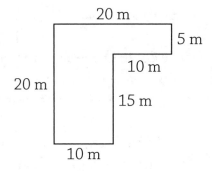

▶ Add the lengths of the six sides of the polygon: $20 + 5 + 10 + 15 + 10 + 20 = 80$.

▶ The perimeter of the polygon is 80 m.

Here's an example of how to find the perimeter of a slightly irregular polygon.

EXAMPLE

▶ Find the perimeter of the polygon.

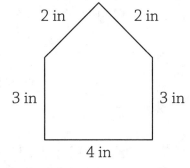

> Add the lengths of the five sides of the polygon: $2 + 2 + 3 + 4 + 3 = 14$.
>
> The perimeter of the polygon is 14 in.

Finding the Circumference of a Circle

Unlike other polygons, a circle doesn't have sides that are straight lines. A **circle** is the set of all points in a plane the same distance from a given point called the **center**. The distance around a circle has a special name called the **circumference** (C).

Finding the circumference of a circle requires knowing the meaning of a number of different terms as well as using multiplication instead of addition. The **diameter** (d) of a circle is a line segment that passes through its center and has both endpoints on the circle. The **radius** (r) of a circle is a line segment that has one endpoint at its center and the second endpoint on the circle. The following diagram shows the names of the various parts of a circle.

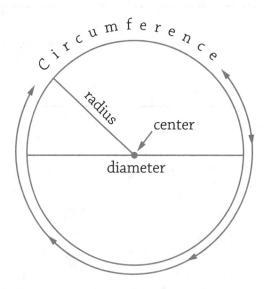

Now, you may still be wondering how this helps us find the circumference of a circle. To answer, we need to know one more piece of information—the meaning of the symbol π (*pi*), which is pronounced the same as the word *pie*. **Pi (π)** is ratio of the circumference of a circle to its diameter. This means that the length of the distance around a circle divided by its diameter is always the same. It doesn't matter how small or large the circle is, its circumference is always π times its diameter!

All we really need to find the circumference of a circle is its diameter: $C = \pi d$. The radius will also do the trick since the diameter equals twice the radius ($d = 2r$). Thus, another way to calculate the circumference of a circle is to use the formula $C = 2\pi r$. Since pi is an **irrational number** (a nonterminating, nonrepeating decimal), both $\frac{22}{7}$ and 3.14 are *approximations* of pi.

The next example shows how to calculate the perimeter of a circle using $\pi = 3.14$.

EXAMPLE

▶ Find the circumference of the circle to the nearest tenths place using $\pi = 3.14$.

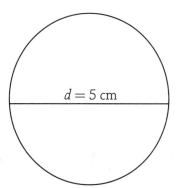

▶ Substitute our values and solve for C.

$C = \pi d$
$C = 3.14 \times 5$
$C = 15.7$ cm

▶ The circumference of the circle to the nearest tenths place is 15.7 cm.

Now, let's look at how to find the circumference of a circle using $\pi = \frac{22}{7}$.

> Find the circumference of the circle to the nearest hundredths place using $\pi = \frac{22}{7}$.

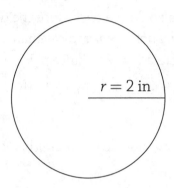

> After we substitute our values, we need to solve for C.
>
> $C = 2\pi r$
> $C = (2 \times 2) \times \left(\frac{22}{7}\right)$
> $C = 4\left(\frac{22}{7}\right)$
> $C = \frac{88}{7}$
> $C = 12.57$

> The circumference of the circle to the nearest hundredths place is 12.57 in.

EXERCISES

EXERCISE 8-1

Find the equivalent measures of these U.S. customary lengths.

1. How many feet are there in 5 yd?

2. How many inches are there in 3 ft?

3. How many yards are there in 2 miles?

4. How many feet are there in 1.5 miles?

EXERCISE 8-2

Find the equivalent measures of these U.S. customary weights.

1. How many pounds do 72 oz of potato salad make?

2. How many ounces are in 6 lb of fish?

3. How many pounds does a 5-T sculpture weigh?

4. How many tons is a 4,500-lb medium size car?

EXERCISE 8-3

Find the equivalent measures of the following U.S. customary units of volume.

1. How many cups are there in 3 pt of milk?

2. How many fluid ounces are there in 2 c of water?

3. How many quarts are there in 3 gal of chicken soup?

4. How many pints are there in 2 qt of orange juice?

EXERCISE 8-4

Find the equivalent measures of the given units of time.

1. How many minutes are there in $3\frac{1}{2}$ hr?

2. How many hours are there in a week?

3. How many seconds are there in 15 min?

4. How many years are there in $2\frac{1}{2}$ centuries?

5. How many weeks are there in a decade?

6. How many years are there in 60 months?

EXERCISE 8-5

Find the equivalent measures of these metric system lengths.

1. How many centimeters long is a 3-m piece of rope?

2. How many millimeters long is a 1.5-m piece of cloth?

3. How many kilometers long is a 7,500-m road?

4. How many decimeters long is a 1.5-m sheet of wrapping paper?

5. How many meters are there in a 5-km race?

EXERCISE 8-6

Say which metric unit you would use to measure the objects given in questions 1 to 3 and then solve the related problems in questions 4 and 5.

1. a grain of salt

2. a textbook

3. a stick of gum

4. If melon has a mass of about 750 g, how many melons would have a mass of 3 kg?

5. If a penny has a mass of 2.5 g, what is its mass in milligrams?

EXERCISE 8-7

Identify the correct metric unit of capacity in the first two questions and then solve the related problems in questions 3 and 4.

1. Which metric unit would you use to measure the capacity of a small spoon?

2. Which metric unit would you use to measure the capacity of a party-size bottle of a favorite soft drink?

3. If a large pot can hold 6,000 ml of soup, how many liters of soup can it hold?

4. If a can of soda holds 350 ml, how many liters would a six-pack of soda contain?

EXERCISE 8-8

Find the perimeter of each polygon.

1.
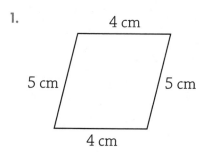

2.

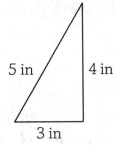

3.

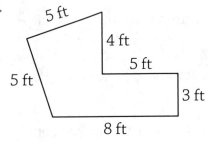

4.

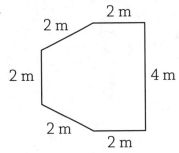

EXERCISE 8-9

Use what we've learned about circumference to answer the next two questions.

1. Find the circumference of this circle to the nearest hundredths place using $\pi = 3.14$.

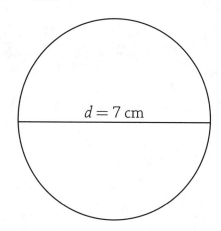

2. Find the circumference of the circle to the nearest hundredths place using $\pi = \dfrac{22}{7}$.

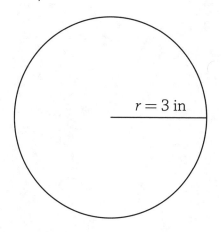

Plane Geometry

MUST KNOW

- A line is a straight, one-dimensional figure that extends forever in both directions. Line segments and rays are made from lines.

- An angle is a figure formed by two rays that meet at a point called the vertex.

- A triangle has three straight sides and three angles that add up to 180°.

- A quadrilateral has four straight sides and four angles that add up to 360°.

- Similarity, congruence, and symmetry are common characteristics of plane figures.

If you look around, you can find lots of examples of two-dimensional figures in the real world. Think of the surface of a sheet of paper, the top of a table, or the side of a door. Notice that we used the word *surface*. That's because these examples refer to only one face of real-world, three-dimensional objects. In fact, though, there are few perfect two-dimensional figures in the natural world. Two-dimensional, or **plane figures,** are ideas created by people.

Identifying Points, Lines, Rays, and Line Segments

Any study of geometry must begin by getting a few ideas under our belt in order to avoid confusion down the road. For example, the idea of a point in geometry is often misunderstood. The confusion arises because when a point is marked on a line or figure, it is represented with a dot (•). In fact, a **point** is simply a location in space. It has no dimension. By convention, a point is named with a capital letter, such as "point A" or "point B."

When you think of a line, you might picture putting a pencil to paper and moving it straight in one direction for a given length, say 3 inches. In geometry, however, a **line** is not a finite line that we can just draw and then be done with, but rather a collection of points that extends without end in opposite directions. To make talking about lines easy, we name them by identifying two points on the line. For example,

$A \quad B$

is named and written as $\overleftrightarrow{AB}$. Notice that, in this written version, the line above the letters A and B has arrows at both ends that point in opposite directions.

A **ray** is a line segment that starts at one endpoint and extends forever in the opposite direction:

Ray CD is written as $\overrightarrow{CD}$, with an arrow pointing off in only one direction.

A **line segment** is part of a line consisting of two endpoints and all the points between them:

The line segment pictured is written as $\overline{EF}$.

> **EXAMPLE**
>
> ▶ Write the name of the figure using the correct symbol.
>
>
>
> ▶ The figure shows a ray because it starts at one endpoint and extends forever in the other direction. The correct way to write this is $\overrightarrow{GH}$.

Let's try a couple of more examples.

> **EXAMPLE**
>
> ▶ Write the name of the figure using the correct symbol.
>
>
>
> ▶ The figure shows a line because it extends forever in both directions. The correct way to write this is $\overleftrightarrow{PQ}$.

EXAMPLE

▶ Write the name of the figure using the correct symbol.

▶ The figure shows a line segment because it has two distinct endpoints. The correct way to write this is $\overline{RS}$.

BTW

When determining how to classify lines, rays, and line segments, be sure to look for dots or arrows at either end.

The direction of a ray, line segment, or line does not change the way it is classified. In other words, a ray is a ray is a ray. The same is true for a line segment and a line.

EXAMPLE

▶ Write the name of the below figure using the correct symbol.

▶ Even though it is the figure is vertical, it shows a line that extends forever in both directions. The correct way to write this is $\overleftrightarrow{XY}$.

Identifying Angles

When two rays come together at a common endpoint, an **angle** is formed. The common endpoint of the angle is called the **vertex**, and the rays are called the **sides** of the angle. The symbol ∠ is used when naming angles.

Angles are named with three capital letters that represent points, or they are identified with numbers:

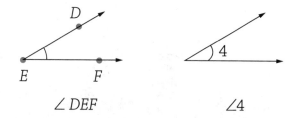

∠ DEF   ∠4

A common way to categorize angles is by the measure of degrees (°) at the angle's vertex:

- A **right angle** has sides that are perpendicular to each other and has a measure of exactly 90°. When the drawing of an angle shows a little square at the vertex, this indicates that it is a right angle.

- An **acute angle** has a measure of less than 90°.

- An **obtuse angle** is greater than 90° but less than 180°.

- A **straight angle** has a measure of exactly 180° and is a straight line.

EXAMPLE

▶ Identify the following angle as *right, acute, obtuse,* or *straight*.

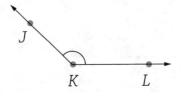

▶ This is an obtuse angle because its measure is greater than 90°.

Looking at the vertex of an angle is the best way to classify it correctly.

EXAMPLE

▶ Identify the angle below as *right, acute, obtuse,* or *straight*.

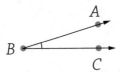

▶ This is an acute angle because its measure is less than 90°.

Sometimes the drawing of angle offers a clear clue as to what type it is.

EXAMPLE

▶ Identify the following angle as *right, acute, obtuse,* or *straight*.

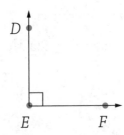

▶ The square at its vertex indicates that this is a right angle with a measure of exactly 90°.

Remember that an angle has two sides and a vertex. What's unusual about the angle shown in the next example?

EXAMPLE

▶ Identify the angle below as *right*, *acute*, *obtuse*, or *straight*.

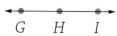

▶ This is a straight angle. Its sides *HG* and *HI* point in opposite directions and its vertex is at *H*.

Finding Angle Measures

Two angles are **complementary** if the sum of their measures is 90°; that is, when taken together, they form a right angle. When the sum of the measures of two angles is 180°, they are called **supplementary** angles:

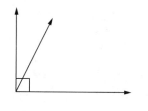   

Complementary Angles   Supplementary Angles

Using what we know about right angles and straight angles can help us estimate other angle measures. Study the following angles to become familiar with "the look" of angle measures:

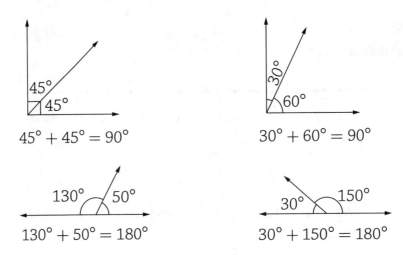

Knowing that a right angle has 90° and a straight angle has 180° makes it easy to determine the measure of an unknown angle with a simple calculation.

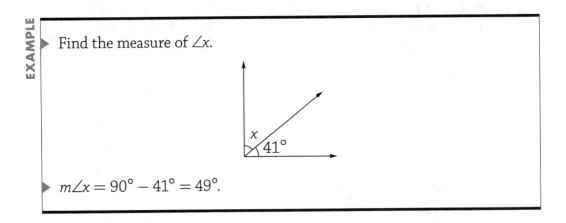

Find the measure of ∠x.

$m\angle x = 90° - 41° = 49°$.

Now, let's try finding the measure of an unknown angle when dealing with supplementary angles.

EXAMPLE

▶ Find the measure of ∠x.

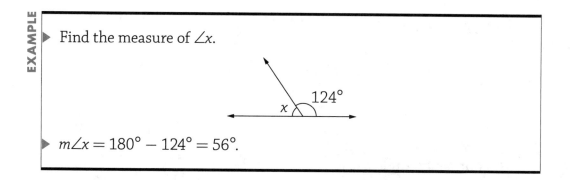

▶ $m\angle x = 180° - 124° = 56°$.

Classifying Triangles

A **triangle** is a closed plane figure, or **polygon**, with three straight sides and three angles. The sum of the angles of a triangle always equals 180°. Each vertex is labeled with a capital letter. We write a triangle's name with the triangle symbol followed by the three vertices, for example, △ABC:

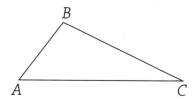

Triangles are classified into two basic groups, by their angles or by their sides:

- An **acute triangle** contains three acute angles.

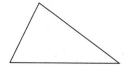

- A **right triangle** contains one right angle.

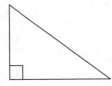

- An **obtuse triangle** has one obtuse angle.

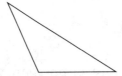

> Identify each triangle as *acute*, *right*, or *obtuse*.

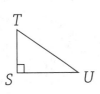 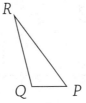 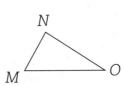

> △TSU is a right triangle since it has one 90° angle. △PQR is an obtuse triangle since it has one angle that is greater than 90°. △MNO is an acute triangle since all three angles are less than 90°.

Different terms are used when we categorize triangles by their sides:

- In an **equilateral triangle** all three sides are the same length.

- In an **isosceles triangle** two sides are the same length.

- In a **scalene triangle** all sides have different lengths.

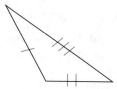

As you can see, we use little tick marks to show that sides are the same or different lengths.

EXAMPLE

▶ Identify each triangle as *equilateral*, *isosceles*, or *scalene*.

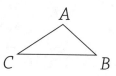 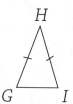 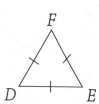

▶ △ABC is a scalene triangle since all three sides are different lengths. △GHI is an isosceles triangle since two of its three sides are the same length. △DEF is an equilateral triangle since all three sides are the same length.

Finding Unknown Angle Measures in Triangles

We can use what we know about the sum of angle measures in a triangle to find an unknown angle measure:

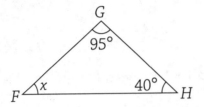

In △ FGH, we can find the measure of ∠x by subtracting the sum of the two known angle measures from 180°:

$$m\angle x = 180° - (95° + 40°)$$
$$= 180° - 135°$$
$$= 45°$$

The same steps can be used to find the missing angle measure in △ JKL.

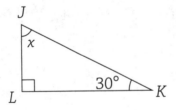

The information given tells us that $m\angle JKL$ is 30°. The square at the vertex of ∠JLK tells us that its measure is 90°. Therefore,

$$m\angle x = 180° - (30° + 90°)$$
$$= 180° - 120°$$
$$= 60°$$

Let's work out a couple of examples.

EXAMPLES

▸ Find the unknown angle x for each triangle:

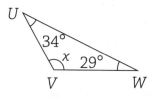 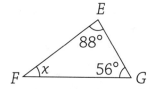

▸ In △UVW, m∠x equals: 180 − (34 + 29) = 180 − 63 = 117°.

▸ In △EFG, m∠x equals: 180 − (88 + 56) = 180 − 144 = 36°.

Classifying Quadrilaterals

A **quadrilateral** is a polygon with four intersecting line segments, called **sides**, and four **vertices** (corners). The sum of the angles of a quadrilateral always equals 360°. Each vertex is labeled with a capital letter. A quadrilateral is represented by the names of its vertices, for example, *EFGH*:

There are several common quadrilaterals that have special features and show up regularly in geometry. It's important to take the time and become familiar with their names and characteristics:

Types of Quadrilaterals		
parallelogram	a quadrilateral with 2 pairs of parallel sides and 2 pairs of equal angles	
rectangle	a parallelogram with 2 sets of equal sides—and 4 right angles	
square	a rectangle with 4 sides of equal length	
rhombus	a parallelogram with 4 sides of equal length and 2 sets of equal angles	
trapezoid	a quadrilateral with exactly one pair of parallel lines	

EXAMPLE

▶ Identify each quadrilateral below as a *rectangle*, *rhombus*, or *trapezoid*.

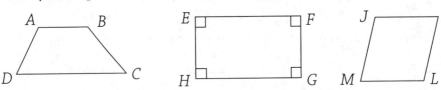

▶ Quadrilateral *ABCD* is a trapezoid since it has only one pair of parallel lines. Quadrilateral *EFGH* is a rectangle since it has four right angles. Quadrilateral *JKLM* is a parallelogram since it has two pairs of parallel lines.

Finding Unknown Angle Measures in Quadrilaterals

Since the sum of the angles of a quadrilateral is 360°, it's easy to figure out the measure of an unknown angle. By definition, every angle in a **square** or **rectangle** is a right angle with a measure of 90°. If the quadrilateral is a **rhombus**, opposite angles always have the same measure and consecutive angles are always supplementary. The simplest way to calculate the measure of a missing angle is to write an equation.

In quadrilateral WXYZ below, we can find the measure of ∠x by subtracting the sum of the three known angle measures from 360°:

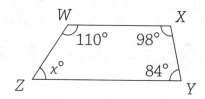

$$m\angle x = 360° - (110° + 98° + 84°)$$
$$= 360° - 292°$$
$$= 68°$$

Let's work out a couple of examples together.

▸ Find the measure of angle x in each figure.

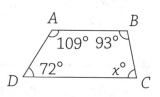 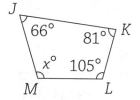

▸ In quadrilateral ABCD, m∠x equals: 360° − (109° + 93° + 72°) = 360° − 274° = 86°.

▸ In quadrilateral JKLM, m∠x equals: 360° − (66° + 81° + 105°) = 360° − 252° = 108°.

Identifying Other Common Polygons

Recall that a polygon is a closed plane figure that is formed by joining three or more line segments at their endpoints to form vertices. You've already studied two types of polygons, triangles and quadrilaterals. Other common polygons include pentagons, hexagons, heptagons, and octagons. Notice that the prefix in each polygon's name refers to the number of sides it has:

Polygon	Regular	Irregular
triangle	△	◁
quadrilateral	□	◁
pentagon	⬠	⬠
hexagon	⬡	⬡
heptagon	⬣	⬣
octagon	⯃	⯃

> **EXAMPLE**
>
> ▶ Identify each polygon below as a *pentagon, hexagon, heptagon,* or *octagon.*
>
>
>
>   A   B   C   D
>
> ▶ Polygon *A* has seven sides and is a heptagon. Polygon *B* has five sides and is a pentagon. Polygon *C* has eight sides and is an octagon. Polygon *D* has six sides and is a hexagon.

Finding Unknown Angle Measures in Other Polygons

We've seen that the sum of the measures of the angles of a triangle equals 180° and that the sum of the measures of the angles of a quadrilateral is always 360°. How do we go about determining the angle measures of the angles of other common polygons? For example, what is the sum of the measures of the angles of a pentagon or a hexagon? Is there a way to determine the measure of each angle in these polygons?

Finding the sum of the measures of the interior angles of a polygon is simple when you use the formula:

$$S = (n - 2)(180°)$$

where S refers to the sum of the angles and n equals the number of sides of the polygon. Therefore, if we want to find the sum of the interior angles of the following polygon, we just count the number of sides and use that number to represent n in the formula.

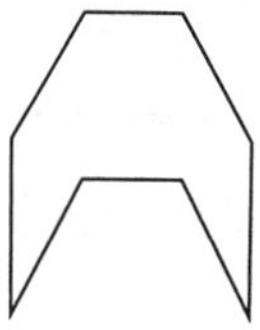

$$\begin{aligned} S &= (n - 2)(180°) \\ &= (8 - 2)(180°) \\ &= (6)(180°) \\ &= 1{,}080° \end{aligned}$$

Notice that the octagon shown above is irregular. Does anything change if you are dealing with a regular polygon—one in which all sides and all angles are the same? The answer is a resounding "No, it doesn't make one bit of a difference!"

However, knowing that a polygon is regular does enable us to find the measure of each of its angles by building on the formula just given. Consider this regular octagon:

From the problem we just worked out, we know that the sum of the interior angles of an octagon is 1,080°. We also know that a regular octagon has eight equal sides and eight equal angles. Therefore, we only need to find: 1,080° ÷ 8 = 135°. So, each angle in a regular octagon has a measure of 135°.

We can use the following formula to find the measure of the interior angle of any regular polygon:

$$m\angle \text{interior angle} = \frac{(n-2)(180°)}{n}$$

> **EXAMPLE**
>
> ▶ Find the sum of the measures of the interior angles of each polygon below.
>
>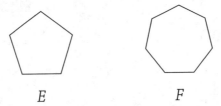
>
> E   F
>
> ▶ Polygon E shows a five-sided figure called a pentagon. Using the formula $S = (n - 2)(180°)$, the sum of the interior angles is: $S = (5 - 2)(180°) = 3(180°) = 540°$.
>
> ▶ Polygon F shows a seven-sided figure called a heptagon. The sum of its interior angles is: $S = (7 - 2)(180°) = 5(180°) = 900°$.

Now, let's find the measure of each interior angle of two regular polygons.

EXAMPLES

▶ Find the measure of an interior angle of each of these regular polygons.

G

H

▶ Polygon G has six sides and is a hexagon. Find the sum of the interior angles: $S = (6 - 2)(180°) = 4(180°) = 720°$. Then, divide the sum of the measures by six: $720° \div 6 = 120°$.

▶ Each angle in a regular hexagon has a measure of 120°.

▶ Polygon H has five sides and is a pentagon. Find the sum of the interior angles: $S = (5 - 2)(180°) = 3(180°) = 540°$. Divide the sum of the interior measures by 5: $540° \div 5 = 108°$.

▶ Each angle in a regular pentagon has a measure of 108°.

Identifying Congruent and Similar Figures

Two two-dimensional figures are **congruent** if they have the same size and shape, that is, if each matching pair of sides and angles has identical measures. Another way of saying this is that congruent figures have identical **corresponding parts** and can be placed exactly on top of each other:

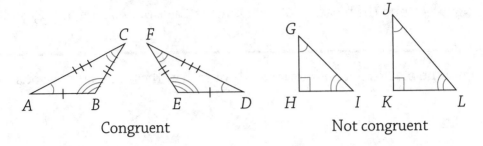

Congruent   Not congruent

Notice that, in the preceding figures, tick marks are used to identify the sides of the triangles that are congruent, and angle marks are used to indicate angles that have the same measure. In $\triangle ABC$ and $\triangle DEF$, side AB is congruent to side DE, side BC is congruent to side EF, and side AC is congruent to side DF. Likewise, each angle of $\triangle ABC$ has a corresponding and congruent angle in $\triangle DEF$. We write that the two triangles are congruent as $\triangle ABC \cong \triangle DEF$.

Although $\triangle GHI$ and $\triangle JKL$ both have the same shape and angle measures, they are not congruent because they are not the same size. The sides of these two triangles are proportional to each other; that is, one is a scale drawing of the other, but they are not identical. If we tried to place $\triangle JKL$ over $\triangle GHI$, it would more than cover it. Figures that are **similar** have the same shape and angle measures but are not the same size.

Knowing that two figures are congruent or similar can help us find an unknown angle measure or unknown length. Look at the following figures:

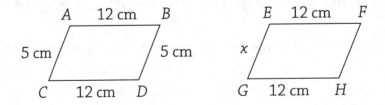

If quadrilaterals $ABCD$ and $EFGH$ are congruent, then side $\overline{EG}$ is congruent with side $\overline{AC}$. Therefore, $\overline{EG}$ equals 5 cm.

In pairs of similar figures, the corresponding angles are always the same measure. However, the measures of their sides are proportional, meaning they share a common ratio. Let's use this fact to find the missing length of one side of one quadrilateral in a pair of similar quadrilaterals.

EXAMPLE

▶ Find the value of x, given the two quadrilaterals below.

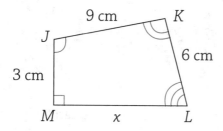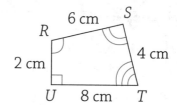

▶ We can see that quadrilaterals JKLM and RSTU are similar figures because the angle marks show that the corresponding angles match up. However, the corresponding sides of the figures are different lengths.

▶ If we look carefully, we can see that corresponding sides ST and KL form a ratio of 4 cm to 6 cm. Sides RS and JK are also corresponding sides, and their lengths are 6 cm to 9 cm.

▶ Notice that these two pairs of corresponding sides actually share the same ratio since: $\frac{4}{6} = \frac{6}{9} = \frac{2}{3}$. This means that every 2 cm in RSTU is 3 cm in JKLM.

▶ Using this ratio, we can set up a proportion to find the length of side LM:

$$\frac{2}{3} = \frac{8}{x}$$
$$2x = 24$$
$$x = 12$$

▶ Side LM, therefore, measures 12 cm.

Identifying Lines of Symmetry

Nature presents us with many beautiful examples of symmetry, from the markings of a butterfly, the petals of a flower, to an individual snowflake:

  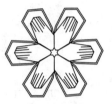

In geometry, a plane figure has **line symmetry** if a line can be drawn that divides the figure into two congruent parts that are mirror images of each other. The line that creates the two mirror images is called the **line of symmetry**. Each of the preceding images can be shown to have a line of symmetry:

Plane figures can have zero, one, two, three, or more lines of symmetry:

 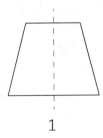   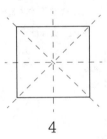

    0   1   2   3   4

Notice that each figure with line symmetry creates mirror images that can be folded along the line of symmetry to form two perfectly overlapping images.

The next figure shows half of a plane image with its line of symmetry. By drawing the reflection of the image over the line of symmetry, we can complete the image:

 IRL Symmetry plays an important part in clothing and fabric design. Clothing designers most often want the left side of a shirt, blouse, pants, or dress to be the same as the right side. The types of patterns printed on fabrics often contain repeating elements that add symmetry and make what we wear interesting to others.

EXAMPLE ▸ Copy each polygon below and draw the line(s) of symmetry. If the figure has no line of symmetry, write "no lines of symmetry."

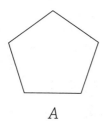  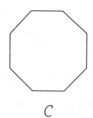

A   B   C

▶ Polygon *A* has five lines of symmetry. Polygon *B* has no lines of symmetry. Polygon *C* has four lines of symmetry:

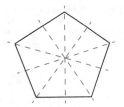 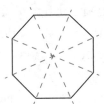

Now, let's try creating a couple of figures using lines of symmetry.

EXAMPLE

▶ Complete the figures below by drawing their reflections over the line of symmetry.

D

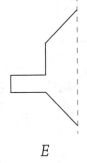

E

▶ Figure *D* looks like the figure below when it is completed by drawing its reflection over the line of symmetry:

▶ Figure *E* looks like the following figure after drawing its reflection over the line of symmetry.

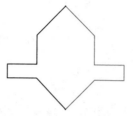

EXERCISES

EXERCISE 9-1

Write the name of each figure using the correct symbol.

1.
   R   S

2. P   Q

3. L   M

4. T   U

EXERCISE 9-2

Identify each angle as right, acute, obtuse, or straight.

1.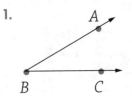

2.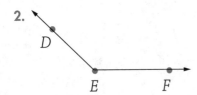

3.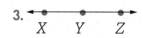

4.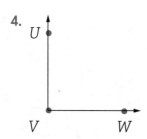

EXERCISE 9-3

Identify each triangle as acute, right, or obtuse.

1.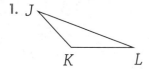

2.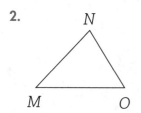

3.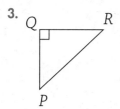

EXERCISE 9-4

Identify each triangle as equilateral, isosceles, or scalene.

1.

2.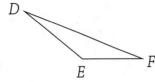

3.

EXERCISE 9-5

Use the accompanying figures to find m∠x.

1.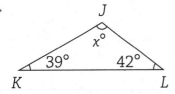

2.

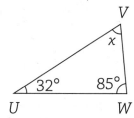

EXERCISE 9-6

Identify each quadrilateral as a parallelogram, rectangle, rhombus, or trapezoid.

1.

2.

3.

4.

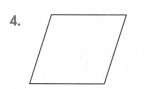

EXERCISE 9-7

Use the accompanying figures to find m∠x.

1.
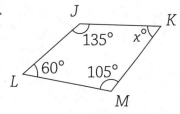

2.
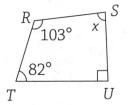

EXERCISE 9-8

Identify each polygon as a pentagon, hexagon, heptagon, or octagon.

1.

2.

3.

4.

EXERCISE 9-9

Find the sum of the measures of the interior angles of each polygon.

1.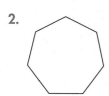

2.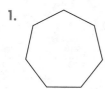

EXERCISE 9-10

Find the measure of an interior angle of each regular polygon to the nearest tenth.

1.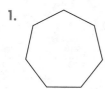

2.

EXERCISE 9-11

Identify each pair of figures as congruent, similar, or neither.

1.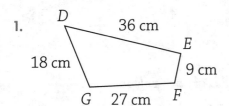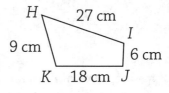

2.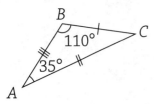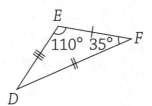

3.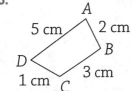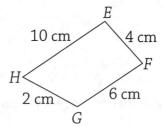

EXERCISE 9-12

Use the accompanying polygons to find m∠x.

1.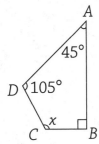

2.

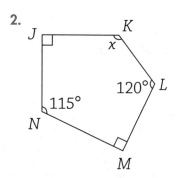

EXERCISE 9-13

For each figure, draw the line(s) of symmetry. If the figure has no line of symmetry, write "none."

1.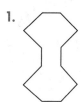

2.

MUST KNOW Math Grade 6

EXERCISE 9-14

Complete each image by drawing its reflection over the line of symmetry.

1.
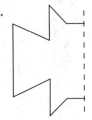

2.

10 Geometry: Area and Volume

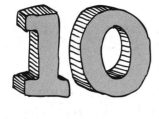

MUST KNOW

⚡ We can find the area of plane figures if we know key measurements such as length, width, height, or radius.

⚡ A rectangular prism has six faces that are rectangles.

⚡ A cylinder is a solid figure with same-size and parallel circles at the top and bottom.

⚡ A cone is a solid figure with a circular base and a curved surface that is joined at a single point.

⚡ A sphere is a solid figure with all the points on its surface the same distance from its center.

⚡ The surface area of a solid figure is the total area of its outer surface. The volume of a solid figure is the amount of space the figure occupies.

nowing how to find the area of plane figures has lots of practical uses in the real world. If we want to carpet a 9-foot by 10-foot room, we'll need to figure out the area of the floor so we know how much carpet to buy. Likewise, knowing how to calculate the volume of solid figures is helpful when you want to determine, say, how much packing material is needed to fill a box.

Area of Plane Figures

The **area** of a **plane figure** is the measure of the number of square units needed to cover the surface of the figure. If we divide the rectangle shown below into square units, we can count the number of square centimeters to find its area:

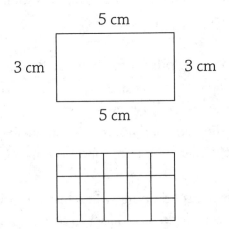

The rectangle contains 15 little squares, each of which is 1 square centimeter (1 cm²). So, the area of the rectangle with sides 3 cm and 5 cm is 15 cm².

Finding the Area of Quadrilaterals

The area of quadrilaterals can be found be multiplying length by width, as in the case of rectangles, or by squaring a side, as with squares:

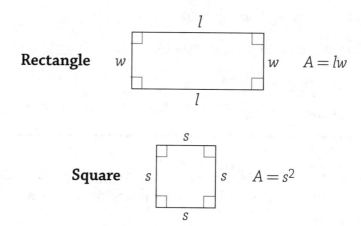

EXAMPLE

▶ Kendis is carpeting his bedroom. The room measures 9 ft by 12 ft. He finds the area of the room to find out how much carpeting he will need. What is the area of his bedroom?

▶ The length of the room is 12 ft and its width is 9 ft. Multiply length by width to find the area.

$A = 12 \text{ ft} \times 9 \text{ ft} = 108 \text{ ft}^2$

▶ The area of the room is 108 ft^2.

Now, let's try solving a problem that involves finding the area of a square.

> **EXAMPLE**
>
> ▶ Jamaica and her daughter Annie are planting a new vegetable garden. The garden will be square and measure 5 yards on each side. What will be the area of their vegetable garden?
>
> ▶ Use the formula for finding the area of a square: $A = s^2$.
>
> $A = (5 \text{ yd})^2 = 5 \text{ yd} \times 5 \text{ yd} = 25 \text{ yd}^2$
>
> ▶ The area of the vegetable garden is 25 yd^2.

Unlike a square and a rectangle, a parallelogram is a quadrilateral that does *not* have right angles. Recall that a parallelogram is a quadrilateral with two pairs of parallel sides:

How do we go about finding the area of a parallelogram? At first glance, the solution would seem unrelated to the way we find the area of rectangle, but that's actually not true.

If we drop a dashed line perpendicular to the base of a parallelogram from the vertex of an angle on the opposite side, we form a triangle. If we move that triangle to the other side of the parallelogram, we can see that a rectangle is formed. The base of the parallelogram functions as the length of the rectangle and its height is equal to the rectangle's width:

CHAPTER 10   Geometry: Area and Volume   209

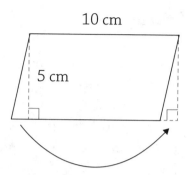

> **BTW**
> Remember that all squares and rectangles are by definition parallelograms. However, not all parallelograms are squares or rectangles. If you keep this distinction in mind, you will always use the correct formula to find the areas of these shapes.

So, while the formula for the area of a square or rectangle is $A = lw$, the formula for area of a parallelogram is $A = bh$. In this case, the area of the parallelogram shown is: $A = 10 \text{ cm} \times 5 \text{ cm} = 50 \text{ cm}^2$.

EXAMPLE

▸ Filipe is making a coffee table with a top in the shape of a parallelogram. The base of the table top is 3 ft and its height is 2 ft. What is the area of the top of the coffee table?

▸ Use the formula for finding the area of a parallelogram: $A = bh$.

   $A = 3 \text{ ft} \times 2 \text{ ft} = 6 \text{ ft}^2$

▸ The area of the coffee table top is 6 ft².

Here's another real-world example.

EXAMPLE

▸ Alisha is creating a new series of paintings. The canvas of each painting is in the shape of a parallelogram that has a base of 10 in and a height of 8 in. What is the area of each canvas?

▸ Use the formula for finding the area of a parallelogram: $A = b \times h$.

   $A = 10 \text{ in} \times 8 \text{ in} = 80 \text{ in}^2$

▸ The area of each canvas is 80 in².

Finding the Area of Triangles

Finding the area of a triangle requires the same terms and concepts we used in finding the area of a parallelogram. Any side of a triangle can be its **base** (*b*). The **height** (*h*) of a triangle is its altitude, or the distance of a line perpendicular to its base from the vertex opposite it. The area of a triangle is simply half that of a rectangle with the same base and height. To find the area of a triangle, then, we use the formula $A = \frac{1}{2}bh$.

If we want to find the area of the triangle shown next, all we have to do is plug the numbers into the formula:

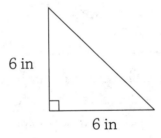

Notice that this is an isosceles right triangle, so its base and its height are both 6 in: $A = \frac{1}{2}(6 \text{ in} \times 6 \text{ in}) = \frac{1}{2}(36 \text{ in}^2) = 18 \text{ in}^2$. So, the triangle has an area of 18 in².

▶ What is the area of the triangle below?

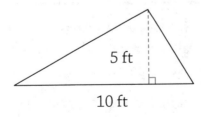

▶ Use the formula for the area of a triangle to find the answer.

$$A = \frac{1}{2}bh$$

$$= \frac{1}{2}(10 \text{ ft} \times 5 \text{ ft})$$

$$= \frac{1}{2}(50 \text{ ft}^2)$$

$$= 25 \text{ ft}^2$$

▶ The triangle has an area of 25 ft².

The next example involves a real-world scenario.

▶ Suppose a pennant for your school has the measurements shown here.

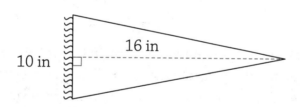

What is the area of the pennant?

▶ Use the formula for the area of a triangle to find the answer.

$$A = \frac{1}{2}bh$$

$$= \frac{1}{2}(10 \text{ in} \times 16 \text{ in})$$

$$= \frac{1}{2}(160 \text{ in}^2)$$

$$= 80 \text{ in}^2$$

▶ The pennant has an area of 80 in².

Finding the Area of Circles

A circular piece of paper has a radius of 4 in. How can we find the area of the following circle?

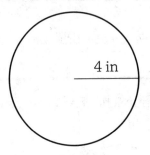

Recall that finding the circumference of a circle involves the use of the special number pi (π). Pi represents the ratio of the diameter of a circle to its circumference. Pi is a constant; that is, no matter how small or large the circle, the relationship between its diameter and its circumference is always the same. Finding the area of a circle also involves the use of pi: $A = \pi r^2$. Notice that, as with the area of other plane figures, the formula deals in square units of measurement. To find the area of any circle, all we have to do is remember the formula, plug in the numbers, and solve. We figure out the area of the above circle this way:

$$A = \pi r^2$$

$$A = \pi (4 \text{ in})^2$$

$$A = (16 \text{ in}^2)\pi$$

$$A = 16 \text{ in}^2 \times 3.14$$

$$A \approx 50.24 \text{ in}^2 \approx 50 \text{ in}^2$$

Notice that in the answer the symbol $\approx$ is used instead of $=$. That's because any calculation involving π is always an approximation. The area of this circle, then, is approximately 50 in².

EXAMPLE

▶ A circular place mat has a radius of 11 cm. What is the area of the place mat to the nearest centimeter?

▶ Use the formula for the area of a circle to find the answer.

$$A = \pi r^2$$

$$A = \pi (11 \text{ cm})^2$$

$$= 121\pi \text{ cm}^2$$

$$= (121 \times 3.14) \text{ cm}^2$$

$$\approx 379.94 \approx 380 \text{ cm}^2$$

▶ The area of the place mat is approximately 380 cm².

Here's a summary of all the formulas we've studied that show how to find the area of plane figures:

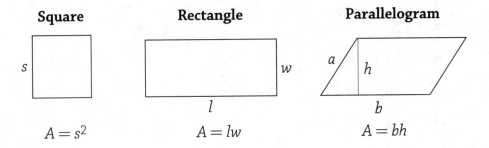

Square
$A = s^2$

Rectangle
$A = lw$

Parallelogram
$A = bh$

Notice that the area of all plane figures is always in square units of measurement such as ft^2, cm^2, yd^2, and mi^2. It's essential to write the exponent next to the unit of measure if you want to write the answer correctly!

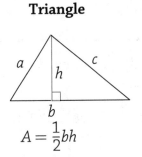

Triangle
$A = \frac{1}{2}bh$

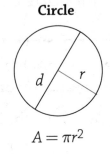

Circle
$A = \pi r^2$

Surface Area of Solid Figures

Solid figures are three-dimensional objects that have length, width, and height. Since we live in a three-dimensional world, it's reasonable to assume that we are surrounded by solid figures—and we are! When the sides of solid figures are polygons, we can use what you know about finding the area of plane figures to find the surface area of the solid.

It's helpful to master a few basic definitions before we begin:

- When polygons form the sides of a solid figure, the sides are called **faces**.

- The lines that are formed when two faces meet are called **edges**.

- Any point where three edges meet is called a **vertex**.

Finding the Surface Area of Rectangular Prisms

We can see how those terms apply to one of the simplest of solid figures—the cube. A **cube** is a solid figure with six congruent sides; that is, sides with the same dimensions. As you'll see, a cube is really a special type of rectangular prism.

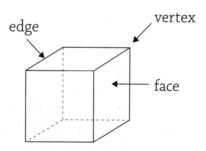

Suppose each edge of a cube is 5 cm long. How could we go about finding the surface area of the entire cube? Well, we already know how to find the area of a square; therefore, we already know how to find the area of one face of the cube. All we have to do, then, is multiply the length of two sides (s^2). Therefore, the area of one of the faces of this cube is: 5 cm × 5 cm = 25 cm^2.

Now, there's just one other step we must take to find the surface area of the entire cube. Since it has six sides, we multiply: 25 cm² × 6. So, the surface area of the cube is 150 cm². The formula for finding the surface area of a cube is:

$$SA = 6s^2$$

EXAMPLE

▶ An artist makes a sculpture of a cube with each edge 3 ft in length. What is the surface area of the cube to the nearest foot?

▶ Use the formula for finding the surface area of a cube: $SA = 6s^2$.

$$SA = 6(3 \text{ ft})^2$$
$$= 6(3 \text{ ft} \times 3 \text{ ft})$$
$$= 6(9 \text{ ft}^2)$$
$$= 54 \text{ ft}^2$$

▶ The surface area of the artist's cube is 54 ft².

A **rectangular prism** is a three-dimensional figure that has six faces, all of which are rectangles. As in the case of a cube, all of the angles in a rectangular prism are right angles. (By now, you've likely figured out that a cube is just a special type of rectangular prism!)

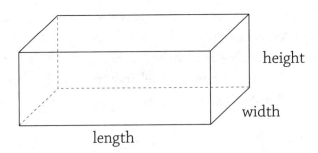

CHAPTER 10 Geometry: Area and Volume

The faces of a rectangular prism form opposite pairs: the top and the bottom are congruent, the sides formed by the length and height are congruent, and so are the faces formed by the width and the height.

The relationships between these opposite faces make finding the surface area of rectangular prism quite logical. What we need to do to find the total surface area is determine the surface area of each type of face and double each.

EXAMPLE

▶ Suppose a box has a length of 20 in, a width of 10 in, and a height of 3 in. What is the surface area of the box?

▶ First, begin by finding the surface area of either the top or bottom by multiplying the length and width and then multiplying by 2.

Top/Bottom: 20 in × 10 in = 200 in^2 × 2 = 400 in^2

▶ Next, find the surface area of the two long sides by multiplying the length and height and then multiplying by 2.

Long Side: 20 in × 3 in = 60 in^2 × 2 = 120 in^2

▶ Then, find the surface area of the two short sides by multiplying the width and height and then multiplying by 2.

Short Side: 10 in × 3 in = 30 in^2 × 2 = 60 in^2

▶ Finally, add all the measures together.

400 in^2 + 120 in^2 + 60 in^2 = 580 in^2

▶ The total surface area of the box is 580 in^2.

The steps we followed in the preceding example give us good insight into the formula for finding the surface area of any rectangular prism:

$$SA = 2lw + 2lh + 2wh$$

> **EXAMPLE**
>
> A stage is in the form of a rectangular prism that is 10 yd long, 4 yd wide, and 5 yd high. What is the surface area of the stage to the nearest yard?
>
> ▶ Use the formula for finding the surface area of a rectangular prism: $SA = 2lw + 2lh + 2wh$.
>
> $$\begin{aligned} SA &= 2(10 \text{ yd} \times 4 \text{ yd}) + 2(10 \text{ yd} \times 5 \text{ yd}) + 2(4 \text{ yd} \times 5 \text{ yd}) \\ &= 2(40 \text{ yd}^2) + 2(50 \text{ yd}^2) + 2(20 \text{ yd}^2) \\ &= 80 \text{ yd}^2 + 100 \text{ yd}^2 + 40 \text{ yd}^2 \\ &= 220 \text{ yd}^2 \end{aligned}$$
>
> ▶ The surface area of the stage is 220 yd².

Finding the Surface Area of Cylinders

A **cylinder** is a three-dimensional solid with two circular bases that are opposite and parallel to each other and a face that connects the two bases.

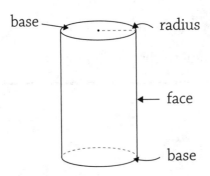

Therefore, the surface area of a cylinder has three parts—the surface area of its two bases and the surface area of its face. We already know how to find the area of a circle, so we are halfway to knowing how to find the surface area of a cylinder.

How can we determine the area of a cylinder's face? If we roll out the face of a cylinder, it forms a rectangle. The width of this rectangle is the same as the height of the cylinder. The length of the unrolled rectangle is exactly the same as the circumference of the circles that form its top and bottom.

Thus, the formula used to find the surface area of a cylinder is:

$$SA = 2\pi rh + 2\pi r^2$$

Which part of the formula gives us the area of the two circles that form the cylinder's bases? Which part of the formula gives us the area of the cylinder's face?

EXAMPLE

▶ A can of coffee has a diameter of 10 cm and a height of 14 cm. What is the surface area of the can to the nearest tenth?

▶ Use the formula for finding the surface area of a cylinder: $SA = 2\pi rh + 2\pi r^2$.

▶ Notice that the problem gives you the diameter of the top and bottom of the can, but the formula requires that you use the radius. Recall that the radius of a circle is half its diameter. So, the correct measure to use is 5 cm.

$$\begin{aligned} SA &= 2(3.14)(5 \text{ cm} \times 14 \text{ cm}) + 2(3.14)(5^2 \text{ cm}^2) \\ &= 2(3.14)(70 \text{ cm}^2) + 2(3.14)(25 \text{ cm}^2) \\ &= 2(219.8 \text{ cm}^2) + 2(78.5 \text{ cm}^2) \\ &= 2(219.8 \text{ cm}^2 + 78.5 \text{ cm}^2) \end{aligned}$$

$$= (2)(298.3 \text{ cm}^2)$$

$$\approx 596.6 \text{ cm}^2$$

▶ The surface area of the can is approximately 596.6 cm².

Finding the Surface Area of Cones

More than likely you know what a cone looks like from eating ice cream. In geometry, a cone is a solid, three-dimensional object with a circular base that is joined by a curved surface to a vertex:

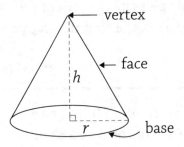

To find the surface area of a cone, we use this formula:

$$SA = \pi rs + \pi r^2$$

In the formula, s refers to "slant height," or the length along the face of the cone from its base to its vertex.

EXAMPLE

▶ A saltshaker in the shape of a cone has a radius of 2 cm at its base and a slant height of 4 cm. What is the surface area of the saltshaker to the nearest hundredth?

▶ Use the formula for finding the surface area of a cone: $SA = \pi rs + \pi r^2$.

$$SA = (3.14 \times 2 \text{ cm} \times 4 \text{ cm}) + (3.14)(2^2 \text{ cm}^2)$$
$$= (3.14)(8 \text{ cm}^2) + (3.14)(4 \text{ cm}^2)$$
$$= (3.14)(8 \text{ cm}^2 + 4 \text{ cm}^2)$$
$$= (3.14)(12 \text{ cm}^2)$$
$$\approx 37.68 \text{ cm}^2$$

▶ The surface area of the saltshaker is approximately 37.68 cm².

A **sphere** is a solid figure in the shape of a ball. It is defined as a solid figure whose surface consists of points all the same distance from the center. The formula for the surface of a sphere is:

$$SA = 4\pi r^2$$

EXAMPLE

▶ A soccer ball has a diameter of approximately 22 cm. What is the surface area of the soccer ball to the nearest centimeter?

▶ Use the formula for finding the surface area of a sphere: $SA = 4\pi r^2$.

▶ Notice that the problem gives you the diameter of the soccer ball, not its radius.

$$SA = 4 \times 3.14 \times 11^2 \text{ cm}^2$$
$$= 4 \times 3.14 \times 121 \text{ cm}^2$$
$$\approx 1{,}520 \text{ cm}^2$$

▶ The surface area of the soccer ball to the nearest centimeter is approximately 1,520 cm².

Volume of Solid Figures

Volume is the amount of space a three-dimensional solid occupies. Volume is always measured in cubic units. Below, we can see how a rectangular prism is made up of individual cubes:

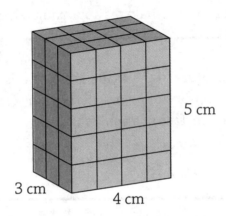

Finding the Volume of Rectangular Prisms

We can find the volume of the above rectangular prism by counting the total number of cubic centimeters. Or we can arrive at the same answer by multiplying the length (l) by the width (w) by the height (h). Therefore, the volume of this rectangular prism is: 4 cm × 3 cm × 5 cm = 60 cm³. The formula for finding the volume of a rectangular prism, including a cube is:

$$V = lwh$$

▶ A cube-shaped box measures 18 cm along each edge. What is the volume of the box to the nearest centimeter?

CHAPTER 10 Geometry: Area and Volume

▸ Use the formula for finding the volume of a rectangular prism: $V = lwh$.

$V = 18 \text{ cm} \times 18 \text{ cm} \times 18 \text{ cm}$

$= 5{,}832 \text{ cm}^3$

▸ The volume of the box to the nearest centimeter is 5,832 cm³.

BTW

The most common mistake made when writing measurements involving area and volume is using the wrong exponent or no exponent at all. When writing about area, be sure to use the exponent 2 to represent two dimensions—for example, cm² and ft². When writing about volume, use the exponent 3 since it represents three dimensions.

Sometimes, solids will have somewhat irregular shapes that are based on rectangular prisms. Consider the following example.

EXAMPLE

▸ A shop owner builds a simple step-up to help customers enter his store. The dimensions of the step-up are:

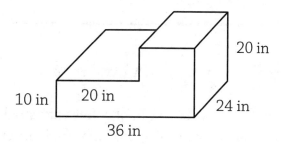

▸ Adjust the diagram so that it shows two rectangular prisms.

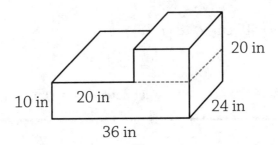

▸ Find the volume of the bottom rectangular prism. Use the formula: $V = lwh$.

$V = 36 \text{ in} \times 24 \text{ in} \times 10 \text{ in}$
$= 8{,}640 \text{ in}^3$

▸ Next, find the volume of the top rectangular prism.

$V = 16 \text{ in} \times 24 \text{ in} \times 10 \text{ in}$
$= 3{,}840 \text{ in}^3$

▸ Add the volume of the two rectangular prisms.

$V = 8{,}640 \text{ in}^3 + 3{,}840 \text{ in}^3 = 12{,}480 \text{ in}^3$

▸ So, the volume of the step-up is $12{,}480 \text{ in}^3$.

Finding the Volume of Cylinders

Earlier in this chapter we saw that a can of coffee with a radius of 5 cm and a height of 14 cm has a surface area of approximately 596.6 cm². How can we find the volume of the can? We first need to find the area of the base of the cylinder (πr^2), and then we multiply this by the height (h) of the cylinder. The formula we use is:

$$V = \pi r^2 h$$

Let's work out the volume of the coffee can.

> **EXAMPLE**
>
> ▶ A can of coffee has a diameter of 10 cm and a height of 14 cm. What is the volume of the can to the nearest centimeter?
>
> ▶ Use the formula for finding the volume of a cylinder: $V = \pi r^2 h$.
>
> ▶ First, find the area of the base of the cylinder. Remember that the radius of the can is half its diameter, or 5 cm.
>
> $$\begin{aligned} A &= \pi r^2 \\ &= (3.14) \times (5^2 \text{ cm}^2) \\ &= (3.14) \times (25 \text{ cm}^2) \\ &= 78.5 \text{ cm}^2 \end{aligned}$$
>
> ▶ Next, find the volume of the can by multiplying the area of the base of the cylinder (πr^2) by its height.
>
> $$\begin{aligned} V &= \pi r^2 h \\ &= (78.5 \text{ cm}^2)(14 \text{ cm}) \\ &= 1{,}099 \text{ cm}^3 \end{aligned}$$
>
> ▶ The volume of the can is approximately 1,099 cm³.

Finding the Volume of Cones

Earlier in the chapter, we determined the surface area of a saltshaker that was in the shape of a cone. To find the volume of a cone, we use the following formula:

$$V = \frac{1}{3}r^2 h$$

> **EXAMPLE**
>
> ▶ A saltshaker in the shape of a cone has a radius of 3 cm at its base and a height of 4 cm. What is the volume of the cone to the nearest tenth?
>
> ▶ Use the formula for finding the surface area of the base of the cone: $A = \pi r^2$.
>
> $$A = \pi r^2 = (3.14)(3^2 \text{ cm}^2)$$
> $$= (3.14)(9 \text{ cm}^2)$$
> $$= 28.26 \text{ cm}^2$$
>
> ▶ Next, plug the area of the base (πr^2) into the formula for the volume of the cone: $V = \frac{1}{3}\pi r^2 h$.
>
> $$V = \frac{1}{3}(28.26 \text{ cm}^2 \times 4 \text{ cm})$$
>
> $$\approx \frac{1}{3}(113.04 \text{ cm}^3)$$
>
> $$\approx 37.7 \text{ cm}^3$$
>
> ▶ The volume of the saltshaker is approximately 37.7 cm³.

Finding the Volume of Spheres

Earlier in this chapter, we determined the surface area of a soccer ball. For a change of pace, let's find the volume of a basketball. To do this, we use the formula for the volume of a sphere:

$$V = \frac{4}{3}\pi r^3$$

> **EXAMPLE**
>
> ▶ A standard basketball has a diameter of 9.5 in. What is the approximate volume of the basketball to the nearest tenth?
>
> ▶ Use the formula for finding the volume of a sphere.
>
> $$V = \frac{4}{3}\pi r^3$$
> $$= \frac{4}{3}(3.14)(4.75^3 \text{ in}^3)$$
> $$\approx \frac{4}{3}(3.14)(107.17 \text{ in}^3)$$
> $$\approx \frac{4}{3}(336.51 \text{ in}^3)$$
> $$\approx 448.73 \text{ in}^3$$
>
> ▶ The volume of the basketball, therefore, is approximately 448.7 in^3.

Here's a summary of the formulas for finding the surface area and volume of common solids:

Cube	Rectangular Prism	Cylinder
$SA = 6s^2$	$SA = 2lw + 2lh + 2wh$	$SA = 2\pi rh + 2\pi r^2$
$V = s^3$	$V = lwh$	$V = \pi r^2 h$

Cone	Sphere
$SA = \pi rs + \pi r^2$	$SA = 4\pi r^2$
$V = \frac{1}{3}\pi r^2 h$	$V = \frac{4}{3}\pi r^3$

CHAPTER 10 Geometry: Area and Volume

EXERCISES

EXERCISE 10-1

Find the surface area of each rectangle.

1. Jason is tiling a wall in his bathroom. The side of each tile measures 5 cm. What is the area of each tile?

2. A theater has a storage basement for scenery, props, and costumes that measures 35 ft by 25 ft. How many square feet is the area of the storage basement?

EXERCISE 10-2

Find the area of each parallelogram.

1. The base of a parallelogram measures 7 in, and its height measures 3 in. What is the area of the parallelogram?

2. What is the area of the parallelogram shown in the figure?

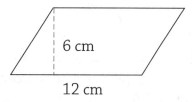

EXERCISE 10-3

Find the area of each triangle.

1.

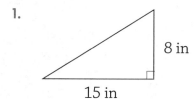

2. A plot of land is in the shape of a triangle with the dimensions shown in the diagram. What is the area of this plot of land?

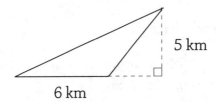

3. A yield sign in the parking lot of EZ Supermarket is in the shape of an equilateral triangle. The sign has sides that measure 18 in each, and its height is 15.6 in. What is the area of the yield sign to the nearest tenth of a square inch?

4. Your bicycling club gives out patches for members who complete the 20-km tour. Each patch is in the shape of an isosceles triangle with the dimensions shown in the diagram. What is the area of the patch to the nearest centimeter?

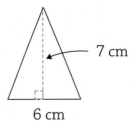

EXERCISE 10-4

Find the area of each circle. Use $\pi = 3.14$.

1. Find the area of the circle to the nearest square centimeter.

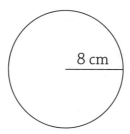

2. The base of a fountain has a diameter of 28 in. What is the area of the fountain's base to the nearest tenth?

3. A circular dartboard has a diameter of 42 cm. What is the area of the surface of the dartboard to the nearest tenth?

4. Sinead is making a bulletin board display that has four circles containing pictures of the student leadership team. Each circle has a radius of 5 in. What is the total area of the four circles to the nearest tenth?

EXERCISE 10-5

Find the surface area of each rectangular prism.

1. What is the surface area of the rectangular prism shown in the diagram?

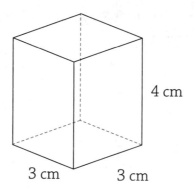

2. What is the surface area of the paper clip box shown in the diagram?

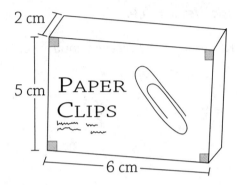

EXERCISE 10-6

Find the surface area of each cylinder.

1. What is the surface area of the cylinder shown in the diagram to the nearest tenth?

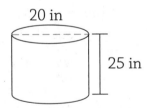

2. A standard D battery has a diameter of about 32 mm and a height of 61.5 mm. If the battery is a simple cylinder, what is the surface area of the battery to the nearest tenth?

EXERCISE 10-7

Find the surface area of each cone.

1. A wax candle is in the shape of a cone. It has a diameter of 10 in and a slant height of 11 in. What is the surface area of the wax candle to the nearest tenth?

2. Find the surface area of the cone shown in the diagram to the nearest inch.

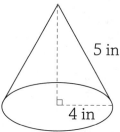

EXERCISE 10-8

Find the surface area of each sphere.

1. Find the surface area of the sphere to the nearest centimeter.

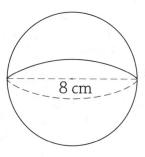

2. A beach ball has a radius of 7 in. What is the surface area of the beach ball to the nearest tenth?

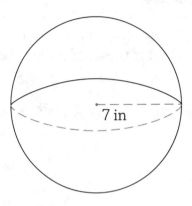

EXERCISE 10-9

Find the volume of each rectangular prism.

1. What is the volume of the rectangular prism in the diagram to the nearest meter?

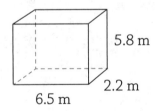

2. A box is 9 ft long, 3 ft wide, and 2 ft high. What is the volume of the box to the nearest foot?

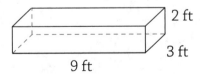

EXERCISE 10-10

Find the volume of each irregular rectangular figure.

1.

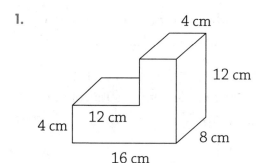

2.

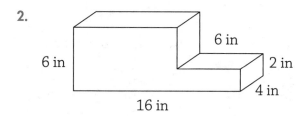

EXERCISE 10-11

Find the volume of each cylinder.

1. A water tank in the shape of a cylinder is 10-m high and has a diameter of 8 m. To the nearest tenth, how many cubic meters of water can the tank hold when it is full?

2. What is the volume of the cylinder shown in this diagram to the nearest meter?

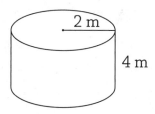

EXERCISE 10-12

Find the volume of each cone.

1. An award in the shape of a cone has a radius of 4 in at its base and a height of 6 in. What is the volume of the cone to the nearest tenth?

2. A traffic cone has a circular base with a diameter of 10 in and stands 18 in high. If the traffic cone is a perfect cylinder, what is its volume to the nearest inch?

EXERCISE 10-13

Find the volume of each sphere.

1. To the nearest tenth, what is the volume of a beach ball with a diameter of 16 in?

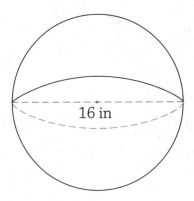

16 in

2. A steel ball in a ball bearing has a diameter of 4 mm. What is the volume of the steel ball to the nearest tenth?

Probability

MUST KNOW

- Probability measures the likelihood that an event will occur. The two kinds of probability are theoretical and experimental.

- A sample space is the set of all possible outcomes in a situation.

- The counting principle states that if event A can happen in m ways and event B in n ways, then events A and B can happen in $m \times n$ ways.

- A combination involves a selection of items from a set in which the order of the items does not matter. In a permutation, the order of the items is important.

Our conversations with relatives and friends often touch on probability. "I heard the weather forecaster say the chance of rain tomorrow is about 50 percent." "There's no way our softball team will win the state championship." "Jamal is certain he will win a basketball scholarship to college." "The odds of me winning the super-lottery are zero!" In each of these examples, what the person is really talking about is the likelihood—or unlikelihood—that certain events will happen. And that's exactly what probability is all about!

Defining Probability

Probability is the area of mathematics that examines the likelihood that an **event**, a particular outcome, will occur. There are actually several types of probability but, for most of us, our introduction to this area is to theoretical probability. **Theoretical probability** assumes that all the possible outcomes are equally likely and are known. It assigns a value based on the number of favorable events to all the possible events.

Simple Probabilities

A simple example of theoretical probability is a coin toss. There are two possible outcomes: heads or tails. Only one event is favorable because we choose either heads or tails. The formula for calculating simple probabilities such as a coin toss is:

$$\text{probability of an event } (P) = \frac{\text{number of favorable outcomes}}{\text{number of possible outcomes}}$$

Therefore, the probability of tossing a heads is $\frac{1}{2}$, and the probability of a tails is also $\frac{1}{2}$.

Simple probabilities such as these can be written in three different ways:

- As a fraction: $\frac{1}{2}$
- As a decimal: 0.50
- As a percent: 50%

> You roll a die with sides labeled 1, 2, 3, 4, 5, and 6. What is the probability you will roll an even number? Write the probability as a percent.
>
>
>
> There are six possible outcomes (1, 2, 3, 4, 5, 6) but only three outcomes are favorable (2, 4, 6).
>
> $$P = \frac{3}{6} = \frac{1}{2}$$
>
> The probability of rolling an even number is 50%.

Let's consider another example using a die.

> What is the probability that you will roll a number greater than 4? Write the probability as a fraction.
>
> There are six possible outcomes (1, 2, 3, 4, 5, 6) but only two outcomes are favorable (5 and 6).

$$P = \frac{2}{6} = \frac{1}{3}$$

▸ The probability of rolling a number greater than 4 is $\frac{1}{3}$.

Sometimes, we have to think carefully about what the problem is asking us to find.

EXAMPLE

▸ If you roll a standard die, what is the probability that you will roll a number that is an integer? Write the probability as a decimal.

▸ There are 6 possible outcomes (1, 2, 3, 4, 5, 6), and all of them are favorable since all the numbers are integers.

$$P = \frac{6}{6} = 1.0$$

▸ The probability of rolling an integer is 1.0. This means that a favorable outcome is *certain* to occur.

Here's an example of a probability at the other extreme.

EXAMPLE

▸ If you roll a standard die, what is the probability that you roll a number greater than 6? Write the probability as a percent.

▸ There are six possible outcomes (1, 2, 3, 4, 5, 6) and none of them are favorable.

$$P = \frac{0}{6} = 0$$

▶ The probability of rolling a number greater than 6 is 0%. This means that a favorable outcome is *impossible* and will never occur.

The next example is a little more complex.

EXAMPLE

▶ A bag contains 10 colored balls. There are 6 red balls, 3 green balls, and 1 orange ball. What is the probability that you will randomly select a green ball from the bag? Write the probability as a percent.

▶ There are ten possible outcomes and three of them are favorable.

$$P = \frac{3}{10} = 0.3$$

▶ The probability of randomly picking a green ball is 30%.

Now, use the set-up from the previous example to find the probability of choosing a different color ball.

EXAMPLE

▶ What is the probability that you will randomly select an orange ball? Write the probability as a percent.

▶ There are ten possible outcomes and one of them is favorable.

$$P = \frac{1}{10} = 0.1$$

▶ The probability of randomly selecting an orange ball is 10%.

> **BTW**
>
> Probability and odds have different meanings. Probability represents the number of times we expect a favorable event to occur divided by the total number of events. The odds of an event are the ratio of the likelihood that the event will occur divided by the likelihood that the event won't occur. An event with 0.60 (60%) probability of occurring has odds of 3 to 2.

Sample Spaces and Tree Diagrams

Counting the number of possible outcomes can be tricky. The set of all possible outcomes in a situation is called the **sample space**. One way to make certain that we haven't missed a possible outcome is to organize the possible outcomes using a **tree diagram** that shows each outcome on a separate branch.

EXAMPLE

▶ A local sandwich shop offers the following choices:

Filling	Cheese	Bread
Tuna	American	Whole wheat
Ham	Cheddar	Rye
Roast beef	Swiss	

Make a tree diagram based on these choices. How many possible sandwich combinations that have a filling, cheese, and bread are there?

▶ To make a tree diagram, first list all the fillings. Add branches that show the different cheeses. Then, add branches that show the breads. Count the number of possible outcomes.

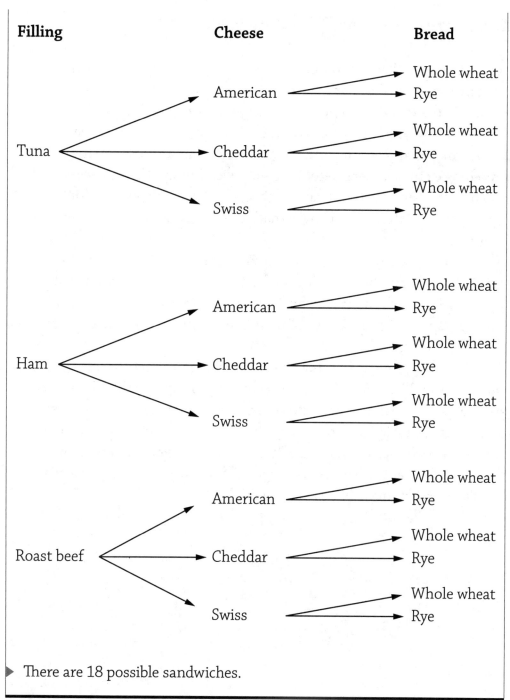

▶ There are 18 possible sandwiches.

Combinations and Permutations

A tree diagram allows us to find all the possible outcomes in a visual way. We can also do this by using the counting principle. According to the **counting principle**, if event A can happen in m ways and event B can happen in n ways, then events A and B can happen in m times n ways.

We can use the counting principle for more than two events. In the preceding example, there were three sandwich fillings, three cheeses, and two breads. Using the counting principle, all we have to do is multiply: $3 \times 3 \times 2 = 18$. Not surprisingly, that's the same answer we got using a tree diagram!

Sometimes, when dealing with a group of items, the order they appear in is not important, but avoiding duplication is essential. A **combination** is a selection of a number of items (objects, people, colors, etc.) from a larger group where order does not matter. Combinations can help us avoid duplication of items.

For example, suppose a school band is looking to select two new members who can play the drums. Four qualified students apply—Alice, Bob, Carla, and Denzel. The bandleader writes the name of each student on a separate piece of paper, folds the papers, puts them in a container, mixes them up, and selects two slips at random. What are the different combinations of two students that the bandleader might choose? Let's fill out a table with everyone else's names next each student:

Choice A Drummer 1	Choice B Drummer 2
Alice	Bob
Alice	Carla
Alice	Denzel
Bob	Alice
Bob	Carla
Bob	Denzel
Carla	Alice
Carla	Bob
Carla	Denzel
Denzel	Alice
Denzel	Bob
Denzel	Carla

At first glance, it would seem that there are 12 possible combinations. However, the order in which the two students are chosen doesn't matter. If Alice and Carla are chosen, that's really the same as choosing Carla and Alice. Getting rid of duplications, the table looks like this:

Choice A Drummer 1	Choice B Drummer 2
Alice	Bob
Alice	Carla
Alice	Denzel
~~Bob~~	~~Alice~~
Bob	Carla
Bob	Denzel
~~Carla~~	~~Alice~~
~~Carla~~	~~Bob~~
Carla	Denzel
~~Denzel~~	~~Alice~~
~~Denzel~~	~~Bob~~
~~Denzel~~	~~Carla~~

So, there are actually only six possible combinations.

> **EXAMPLE**
>
> ▶ Gino's Pizza Shop offers two toppings with any medium-size pie. Toppings include mushrooms, pepperoni, peppers, and onions. How many possible pizza-topping pairings can a customer make? Show your work in a table.
>
> ▶ For our table, let's create columns for the four ingredients as well as a column for the resulting pairs of combinations. Then put Xs in each possible place in the first four columns that will make pairs.
>
> | Mushrooms | Pepperoni | Peppers | Onions | Combination toppings |
> |---|---|---|---|---|
> | X | X | | | Mushrooms/Pepperoni |
> | X | | X | | Mushrooms/Peppers |
> | X | | | X | Mushrooms/Onions |
> | | X | X | | Pepperoni/Peppers |
> | | X | | X | Pepperoni/Onions |
> | | | X | X | Peppers/Onions |
>
> ▶ There are 6 possible pizza-topping pairs.

A **permutation** is an arrangement of items, or outcomes, in which the order *is* important. For example, suppose we wanted to find out how many two-digit numbers we can make using two different digits from 1, 3, 5, 7, and 9.

We can tackle this problem by making an organized list of all possible outcomes.

Numbers beginning with 1: 13, 15, 17, 19
Numbers beginning with 3: 31, 35, 37, 39
Numbers beginning with 5: 51, 53, 57, 59
Numbers beginning with 7: 71, 73, 75, 79
Numbers beginning with 9: 91, 93, 95, 97

So, 20 two-digit numbers are possible permutations.

Making lists of outcomes, however, can be time-consuming and error-producing. A better way is to use factorials and some simple calculations. Recall that a factorial such as 5! represents: $5 \times 4 \times 3 \times 2 \times 1 = 120$.

Suppose five students are competing in a song-writing contest. Only three of the five can win medals. How many different permutations of students can win a gold, a silver, and a bronze medal?

We already saw that with 5! there are 120 possible outcomes. However, here only the top three songs will win. The question then becomes, "How do we factor this detail into our solution?" The answer: Simply take only the first three numbers in the factorial: $5 \times 4 \times 3 = 60$. In all, there are 60 possible ways to arrange the three winners out of a field of five songs.

EXAMPLE

▸ There are seven semifinalists in a student writing competition for a college scholarship. The top three will move on to the finalist stage. How many possible arrangements of the finalists are there?

▸ If there are seven semifinalists, then: $7! = 7 \times 6 \times 5 \times 4 \times 3 \times 2 \times 1$. However, you are only interested in the first three semifinalists, whoever they may be.

▸ Multiply the first three terms of the factorial: $7 \times 6 \times 5 = 210$.

▸ There are 210 possible arrangements for the three finalist spots.

Probability of Independent Events

Two events are **independent** when the outcome of the first event does not affect the outcome of the second event. For example, you flip a coin *and* spin a spinner with evenly spaced numbers 1 through 6. What's the probability that you'll flip a heads on the coin and the number 6 on the spinner?

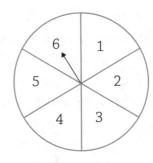

Since the outcome of the coin flip does not affect the number you spin, the two events are definitely independent of each other. One way to answer the preceding question is to make a list of all possible outcomes for both events:

Event 1 Coin Flip	Event 2 Spinner
Heads	1
Heads	2
Heads	3
Heads	4
Heads	5
Heads	**6**
Tails	1
Tails	2
Tails	3
Tails	4
Tails	5
Tails	6

By looking at the rows in the table, it's easy to see that there are 12 outcomes in all. Only one of the twelve is the favorable event you are hoping for. Therefore, the probability of flipping heads and spinning number 6 is $\frac{1}{12}$.

Now, suppose you wanted to find the probability of flipping a tails and spinning an odd number. Obviously, you can look for favorable outcomes using the same table as in the situation just described, and you'll find there are more of them than just one. Again, there are two possible outcomes when flipping a coin: heads and tails. However, there are three ways to spin an odd number: 1, 3, and 5.

Event 1 Coin Flip	Event 2 Spinner
Heads	1
Heads	2
Heads	3
Heads	4
Heads	5
Heads	6
Tails	**1**
Tails	2
Tails	**3**
Tails	4
Tails	**5**
Tails	6

In this case, 3 of the 12 possible outcomes are favorable. So, the probability of flipping a tails and spinning an odd number is: $\frac{3}{12} = \frac{1}{4}$.

We don't have to make a long list to find the probability of two independent events occurring. Instead, we can use the formula: $P(A \text{ and } B) = P(A) \times P(B)$.

EXAMPLE

▶ You flip a coin. Then, you spin a spinner with five equal segments labeled A, B, C, D, and E. What is the probability that you will flip a heads and spin a vowel? Create a table and write the probability as a percent.

▶ The table for our scenario involves putting the letters against both "Heads" and "Tails." We can put the two desired outcomes in bold.

| Event 1 | Event 2 |
Coin Flip	Spinner
Heads	**A**
Heads	B
Heads	C
Heads	D
Heads	**E**
Tails	A
Tails	B
Tails	C
Tails	D
Tails	E

▶ Out of a total of 10 outcomes, 2 are favorable.

$$\frac{2}{10} = \frac{1}{5} = 0.20 = 20\%$$

▶ The probability of flipping a heads and spinning a vowel is 20%.

Here's another example, but with larger numbers.

EXAMPLE

You spin a spinner with 10 equal segments labeled A, B, C, D, E, F, G, H, I, and J. Then you throw a die labeled 1, 2, 3, 4, 5, and 6. What is the probability that you will spin a vowel and roll a number greater than 4? Use the formula for finding the probability of independent events. Write the probability as a decimal.

▶ The probability of spinning a vowel is 3 of 10 possible outcomes. The probability of rolling a number greater than 4 is 2 out of 6.

▶ The probability of spinning a vowel and rolling a number greater than 4 is:

$$P(A \text{ and } B) = P(A) \times P(B)$$

$$= \frac{3}{10} \times \frac{2}{6} = \frac{6}{60} = 0.10$$

▸ The probability of spinning a vowel and tossing a number greater than 4 is 0.10.

Sometimes, a situation will involve more than two events.

EXAMPLE

▸ You have three pairs of pants: black, blue, and brown. You have four T-shirts: one white, one blue, one yellow, and one red. You have four pairs of socks: one black, two blue, and one brown. You randomly select a pair of pants, a T-shirt, and a pair of socks without looking. What is the probability that you will choose blue pants, a blue T-shirt, and a blue pair of socks? Use a formula to solve. Write your answer as a fraction.

▸ The probability that you will pick a pair of blue pants is one out of three, or $\frac{1}{3}$. The probability that you will pick a blue T-shirt is one out of four, or $\frac{1}{4}$. The probability that you will pick blue socks is two out of four, or $\frac{1}{2}$.

▸ Since the events are independent, the probability is:

$$\frac{1}{3} \times \frac{1}{4} \times \frac{1}{2} = \frac{1}{24}$$

▸ The probability that you will pick blue pants, a blue T-shirt, and blue socks is $\frac{1}{24}$.

Experimental Probability

All the probability examples that we've studied so far in this chapter have involved theoretical probability. Recall that theoretical probability is calculated based on the number of favorable events occurring in relation to all possible events. **Experimental probability** is based on real-world experiences or observations. It is calculated based on the number of times an event occurs compared to the number of trials performed. (We can look at the number of "trials" as the number of tries we give something.)

In theoretical probability, the probability of tossing a heads is $\frac{1}{2}$, or 50%, and the probability of tossing a tails is the same. However, we don't actually toss a real coin. In fact, theoretical probabilities don't guarantee that if we toss a coin 10 times, we'll get heads 5 times and tails 5 times. In an *experiment* involving 10 trial tosses, we might toss 4 heads and 6 tails or 7 heads and 3 tails. So, the theoretical probability and the experimental probability don't necessarily match up.

In the real world, there are many situations in which we use experimental probability. For instance, car companies routinely check the brake systems of newly manufactured cars. Suppose a car company tests the brakes on 100 new model cars and finds that 4 of them have faulty brakes. The experimental probability that a car of this model will have brake problems can be easily calculated:

$$P = \frac{4}{100} = 0.04$$

Based on this data, how many new model cars would the company expect to have brake problems if it made 10,000? To find out, simply multiply: 10,000 × 0.04 = 400. The company, then, would expect that about 400 out of the 10,000 newly manufactured cars would have brake problems.

CHAPTER 11 Probability 253

> **EXAMPLE**
>
> ▸ Jeff tosses a bottle cap 50 times. It lands up 28 times and down 22 times. What is the probability that, on the next toss Jeff makes, the cap will land up?
>
> ▸ Calculate the experimental probability of the cap landing heads up based on the data from Jeff's tosses.
>
> $$P = \frac{28}{50} = 0.56 = 56\%$$
>
> ▸ There's a 56% chance the next cap Jeff tosses will land up.

Experimental probability is often used to help businesses make sound decisions for the future.

> **EXAMPLE**
>
> ▸ Sarah Lester runs Ye Olde Bakery Shoppe. She wants to figure out how many of each flavor of cake and pie to make during the upcoming Thanksgiving week. Last year, she sold 120 pies and cakes as shown in the following table.
>
> | | |
> |---|---|
> | Pecan pie | 30 |
> | Pumpkin pie | 35 |
> | Carrot cake | 21 |
> | Black velvet cake | 22 |
> | Orange cake | 12 |
>
> ▸ This year Sarah wants to increase her sales and plans to make 160 pies and cakes for sale during Thanksgiving week. Based on last year's sales, how many pecan pies should Sarah plan to sell?

- Last year, the bakery shop sold 30 pecan pies out of 120 pies and cakes. Calculate the probability of selling a pecan pie based on last year's sales.

$$P = \frac{30}{120} = 0.25 = 25\%$$

- To find out the number of pecan pies she should make, multiply 25% by the total number of pies: $160 \times 0.25 = 40$.

- Sarah should make 40 pecan pies this year.

IRL Drug manufacturers use experimental probability to evaluate for the effectiveness of new medications.

EXERCISES

EXERCISE 11-1

Find the probability of each event.

1. A bag contains 10 marbles. There are four red marbles, two blue marbles, and four green marbles. If a marble is picked at random from the bag, what is the probability that it will be a blue marble? Express the probability as a percent.

2. A spinner has eight equal sections labeled *A, B, C, D, E, F, G,* and *H*. What is the probability that the spinner will land on consonant? Write the probability as a decimal.

3. Eight tiles with letters are laid in a row to form the word *LAUGHTER*. The tiles are placed in a brown paper bag. One tile is randomly selected. What is the probability that the tile will have the letter *I* on it? Write the probability as a whole number.

4. A regular dodecahedron is a three-dimensional shape that has 12 equal pentagonal faces. If the faces are numbered from 1 to 12, what is the probability when rolled that the dodecahedron will show a number greater than 8 on its top face? Write the probability as a fraction.

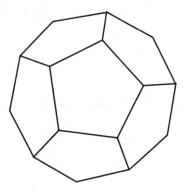

EXERCISE 11-2

Make a tree diagram to show all possible outcomes of each situation.

1. A penny and a nickel are tossed. Make a tree diagram that shows all possible heads and tails outcomes in the sample space. How many possible outcomes are there?

2. You have three kinds of wrapping paper: gold, silver, and blue; three kinds of ribbon: white, pink, and cream; and two kinds of gift tags: square and round. Make a tree diagram to show all possible outcomes of wrapping paper, ribbon, and gift tags. How many possible outcomes are there?

EXERCISE 11-3

Identify all possible combinations of the situation described.

1. The school chorus has two positions open for tenors. Three boys—Abe, Ben, and Charles—apply to fill these spots. What are the different combinations of two tenors that the chorus master might choose? Make a table to show all combinations of two tenors.

2. For lunch, the school cafeteria is offering four main courses and two drinks. If lunch consists of one main course and one drink, how many possible combinations of lunches are there? Make a tree diagram to show all combinations of lunches.

EXERCISE 11-4

Identify all possible permutations of the situation described.

1. How many two-letter permutations can be formed from the letters in *BAT*? Show your answer by making a list.

2. Pete's Hardware needs two employees to work on Saturday. One will work the morning shift and the other the afternoon shift. Pete can select from four workers: Deb, Ed, Fran, and George. How many different permutations of two workers can be selected to work on Saturday? Show your answer by making a list.

EXERCISE 11-5

Find the probability of each event.

1. Rachel and Suzie both go to a see a new movie on Saturday afternoon. They can chose between a sci-fi film and a romantic comedy. Assuming their choices are independent events, what is the probability that they both go to the sci-fi movie? Write the answer as a fraction. Explain your answer.

2. If you roll two standard dice, what is the probability that you will roll two 6s?

3. A bag contains five white balls and five red balls. A second bag contains two black balls and three yellow balls. If one ball is selected randomly from each bag, what is the probability that the two balls you chose are white and black?

4. A spinner is divided into four equal sections. Each section is a different color: red, blue, green, and orange. You make two spins. What is the probability that both spins are orange?

EXERCISE 11-6

Use the tables provided to find the probability of each event indicated.

1. A random survey at the Reynard County schools asked students to choose their favorite fruit drink. The table shows the results of the survey. What is the probability that the next randomly selected student will say that apple juice is his or her favorite fruit drink? Write the probability as a decimal.

Favorite Fruit Juices	Number of Students
Orange	170
Apple	145
Grape	103

2. A spinner is divided into eight equal sections numbered from 1 to 8. Felipe spins 40 times. The table summarizes the results of his trials. What is the experimental probability that the next time Felipe spins, the spinner will land on 4? Write the probability as a decimal.

Result	1	2	3	4	5	6	7	8
Number of Spins	4	5	6	5	4	6	5	5

Data and Statistics

MUST KNOW

- Statistics involves the collection and analysis of data.

- Measures of central tendency include the mean, the median, and the mode.

- Pictographs, bar graphs, circle graphs, and line graphs are common forms of data display.

Statistics plays an essential role in many aspects of everyday life. By relying on the collection and analysis of past data, meteorologists are better able to predict the course of hurricanes and other future weather patterns. Statistics help businesses analyze the success or failure of past decisions and provide data to guide them in making decisions concerning future investments. Statistics plays an essential role in medicine, from the diagnoses of illnesses to evaluations of the effectiveness of new medications.

Defining Statistics

Statistics is a branch of mathematics that studies data, including the methods used to collect, organize, interpret, and represent this data. **Data** are units of information usually in the form of quantitative facts (numbers). **Analysis** is the process of inspecting and evaluating data according to key characteristics.

Measures of Central Tendency

Measures of central tendency are statistical values that we use to summarize a set of data by identifying the set's center. The three most common measures of central tendency are the mean, median, and mode. Let's look at what these measures are and what they reveal about a set of data.

Fifteen students in Mrs. Rivera's math class take their midterm exams. Here are all 15 scores arranged from least to greatest:

60, 65, 68, 72, 76, 80, 80, 84, 85, 87, 89, 90, 92, 93, 94

Based on all the test results, the average, called the **mean**, score is 81. We can find the mean by dividing the sum of data by the number of items in the data set. If we add all the test scores together, they equal 1,215 points. To find the mean, simply divide the sum 1,215 by 15, since this is the number of test scores: $1{,}215 \div 15 = 81\%$.

The **median** of a data set is the middle number when the items are arranged from least to greatest:

$$60, 65, 68, 72, 76, 80, 80, \mathbf{84}, 85, 87, 89, 90, 92, 93, 94$$

Since this data set has 15 items, the eighth number in the data array represents the median, or 84%.

The **mode** is the number that occurs most frequently in a set of data. Since the number 80 appears twice in our data set and all the other scores appear just once, 80 is the mode of the data set.

Sometimes it is helpful to identify the **range**, or spread, in the value of items in a set of data. The range is the difference between the greatest value and the smallest value. In our data set, the range is 34, since: $94 - 60 = 34$.

EXAMPLE

▶ You are raising money for a library book drive at your school. You collect 13 pledges in the amounts shown below. What is the average dollar pledge?

$$\$5, \$5, \$7, \$8, \$10, \$10, \$10, \$15, \$15, \$20, \$20, \$20, \$50$$

▶ There are 13 pledges, which add up to $195. To find the mean: $\$195 \div 13 = \$15$. So, the mean pledge is $15.

The next example uses the same set of data from the library book drive, but this time we are asked to find a different measure of central tendency.

> **EXAMPLE**
>
> ▸ What is the median dollar pledge from the library book drive?
>
> ▸ Identify the middle number in the data.
>
> $5, $5, $7, $8, $10, $10, **$10**, $15, $15, $20, $20, $20, $50
>
> ▸ The median pledge is $10.

Now, let's use the same data set to find the mode.

> **EXAMPLE**
>
> ▸ What are the modes of the data set from the library book drive? (Some data sets have more than one mode.)
>
> ▸ Identify the numbers that appears most frequently in the data.
>
> $5, $5, $7, $8, **$10, $10, $10**, $15, $15, **$20, $20, $20**, $50
>
> ▸ The $10 and $20 pledges are the modes of this data set.

The next example shows how to find the range of this data set.

> **EXAMPLE**
>
> ▸ What is the range of the data set from the library book drive?

> Identify the least number and greatest number that appear in the data.
>
> **$5**, $5, $7, $8, $10, $10, $10, $15, $15, $20, $20, $20, **$50**
>
> The range of this data set is $45 since: $50 − $5 = $45.

Pictographs

A **pictograph** displays data using pictures. For example, the following pictograph shows the favorite type of music of 8th grade students in a middle school survey:

Students' Favorite Type of Music	
Rock	♪♪♪♪♪
Rap/Hip-Hop	♪♪♪♪
Pop	♪♪
Country	♪♪
Other	♪

Each ♪ represents 10 students.

One of the first things to look at when reading a pictograph is the key, which usually appears below the graph. In this case, the key indicates that each musical note symbol ♪ stands for 10 student responses.

How many students in the survey prefer rock? Since there are five ♪ symbols in the row labeled Rock, we can find out by multiplying: 5 × 10 = 50. So, 50 of the students surveyed said rock was their favorite type of music.

We can use a pictograph to make calculations about the data represented. How many more students preferred rock music to country music in this survey? We already know that 50 students in the survey preferred rock. Since there are two musical notes next to Country Music, this indicates that 20 of the students interviewed preferred this type of music. Now, we're ready to take the final step by subtracting: $50 - 20 = 30$. Therefore, 30 more students chose rock as their favorite than chose country music.

> In a survey, 2,000 people chose their favorite ice cream flavor. Using the data in the pictograph, how many people selected vanilla?

Favorite Flavors of Ice Cream

Vanilla	🍦🍦🍦🍦🍦🍦🍦
Chocolate	🍦🍦🍦🍦🍦
Strawberry	🍦🍦🍦🍦
Cookies 'n' Cream	🍦🍦
Pistachio	🍦
Other	🍦

Each 🍦 represents 100 people.

> Count the number of ice cream cones next to Vanilla.
> Multiply: $7 \times 100 = 700$.
> 700 people surveyed chose vanilla as their favorite favor of ice cream.

CHAPTER 12 Data and Statistics 265

Now, let's use data from the same pictograph to make a comparison.

> **EXAMPLE**
>
> ▶ How many more people chose chocolate as their favorite flavor than chose cookies 'n' cream?
>
> ▶ First, count the number of cones next to Chocolate and multiply by 100: $5 \times 100 = 500$.
>
> ▶ Next, count the number of cones next to Cookies 'n' Cream and multiply by 100: $2 \times 100 = 200$.
>
> ▶ Finally, subtract: $500 - 200 = 300$.
>
> ▶ 300 more people chose chocolate as their favorite favor of ice cream than chose cookies 'n' cream.

Bar Graphs

A **bar graph** compares quantities by using vertical or horizontal bars of different lengths to represent data. When reading a vertical bar graph, we look at the top of each bar and then read the value that's listed on the left. With horizontal bar graphs, the scale of values is along the bottom of the graph. The following bar graph shows the life expectancies of several types of farm animals:

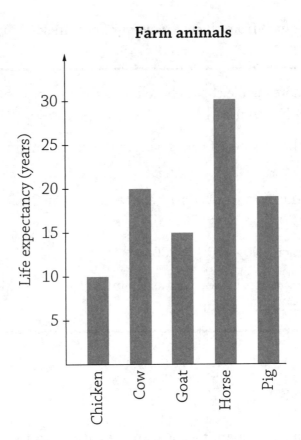

Using the bar graph, which farm animal has the shortest life expectancy? We can answer this question at a glance. Since the bar for chicken is the shortest, that's the correct answer.

Now, use the bar graph to identify which animals have life expectancies of 20 years or greater. Locate 20 years on the vertical scale. Then, look at the heights of the bars. The two farm animals whose bars reach to 20 or higher are the cow and horse.

We can also do calculations using a bar graph. For example, based on the average life expectancies shown, about how many years longer does a pig live than a chicken? To do this, first locate the data for the pig and the chicken by reading the horizontal scale. Then, check the value of the bar and subtract. A pig lives an average of 18 years and a chicken lives an average of

10 years. So, by subtracting 10 from 18, we see that, on average, a pig lives 8 years longer than a chicken.

Now, let's try reading a horizontal bar graph.

EXAMPLE

▶ The bar graph below shows the shirt colors of 100 students selected at random. Using the data in the bar graph, how many students had white shirts?

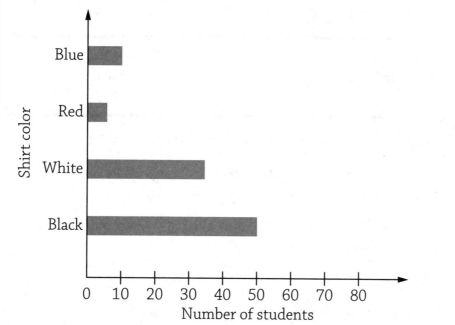

▶ Read the vertical scale to find the bar that represents students with white shirts.

▶ Look at the right-hand side of the white shirt bar and then at the numbers on the horizontal scale.

▶ 35 of the students surveyed wore white shirts.

The next example asks about the data from the same graph about shirt colors.

> **EXAMPLE**
>
> ▶ What color shirt did the fewest number of students wear?
>
> ▶ Look at the lengths of the bars to find the shortest one and read the label below it on the horizontal scale.
>
> ▶ 5 of the students surveyed wore red shirts.

Let's use the shirt graph one more time, to make a comparison.

> **EXAMPLE**
>
> ▶ How many more students wore black shirts than white shirts?
>
> ▶ Look at the vertical scale to locate the bars that represent students with black shirts and with white shirts.
>
> ▶ Look at the right-hand sides of the black shirt and white shirt bars and then at the numbers on the horizontal scale.
>
> ▶ Subtract: $50 - 35 = 15$.
>
> ▶ 15 more students have black shirts than white shirts.

Circle Graphs

A **circle graph**, also called a pie chart, displays data in sections that correspond in area to the relative size of the quantities represented. The entire circle in a circle graph represents 100% of all the data, and each section represents a part of the whole. The larger the area of a section of a circle graph, the greater the quantity it represents in relation to the whole.

For example, the following pie chart shows student choices for their favorite type of movie:

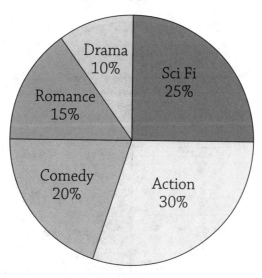

Favorites Types of Movies

How do you know which type of movie is the students' favorite? Since the section labeled Action is the largest, it represents the students' favorite type of movie. Which is the students' least favorite type of movie? The section labeled Drama is the smallest, so it is the students' least favorite type of movie.

BTW

When creating a circle graph, remember that a circle has 360 degrees. If a section of the circle is supposed to represent 25%, then it should have an interior angle of 90°, since 0.25 of 360° equals 90°.

EXAMPLE

▸ Look at the circle graph showing favorite pizza toppings below. Which pizza topping is the students' least favorite?

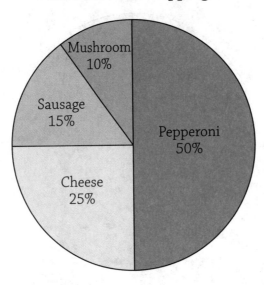

Favorite Pizza Toppings

▸ Look at the circle and find the smallest section.

▸ Mushrooms are the students' least favorite topping.

We can make comparisons based on the size of sections of a circle graph.

EXAMPLE

▸ Use the circle graph showing favorite pizza toppings from the previous example. Which pizza topping do students like twice as much as cheese?

▸ Look at the circle and find the section labeled Cheese. It represents about 25%. Now, find the section that is twice as large as the Cheese section.

▸ Pepperoni is twice as popular as the students' favorite topping as cheese.

Line Graphs

A **line graph** displays changes in data over time using points connected by line segments. When creating a line graph, the points representing specific data are plotted first and then connected by lines. Because of this procedure, a line graph makes **trends**, or changes in the data, easy to see. The following line graph shows the percent of the U.S. workforce employed in agricultural jobs from the year 1900 to 2000:

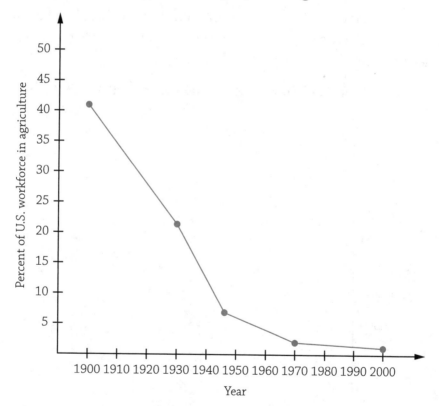

The graph gives us the opportunity to ask a number of questions about the agriculture workforce. For example, between the years 1900 and 1930, how much did the number of agricultural workers decline?

First, read the dates along the horizontal axis of the graph to find 1900. Next, move up to find the point on the line that represents the data for this year. Finally, move left from this point to the vertical axis to find the percent of the workforce employed in agriculture. Now, take the same steps to locate the data for 1930. By subtracting the data (41% − 21%), we see that the percentage of workers employed in agriculture declined about 20% between 1900 and 1930.

Let's work through another graph together.

EXAMPLE

▶ See the line graph below showing how much money was raised during Book Sale Week for each year between 2009 and 2019. How much money was raised during the 2018 Book Sale Week?

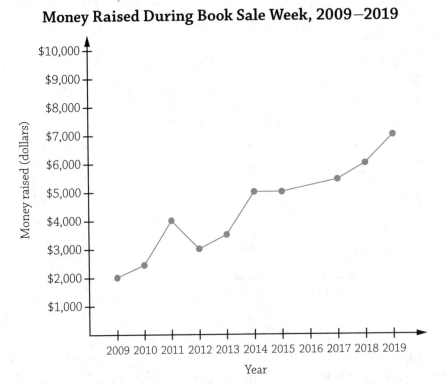

▶ Read the horizontal axis to find the year 2018.

CHAPTER 12   Data and Statistics   **273**

- Move up to find the point related to the year.
- Move left to read the dollars raised on the vertical axis.
- $6,000 was raised during the 2018 Book Sale Week.

When locating data on a line graph, it's important to identify the location of a given point compared to other points on the line.

EXAMPLE

- Use the line graph for the Book Sale Week from the previous example. During which year did the amount of money raised decline?
- Look for a place where the line stops rising and drops.
- Move down to the horizontal axis to find out in which year this occurred.
- The 2012 Book Sale Week showed a decline in the amount of money raised.

It's also important to understand what it means when a section of a line graph does not change.

EXAMPLE

- Use the line graph for the Book Sale Week. During which two years did the amount of money raised stay the same?
- Look for a place where the line neither rises nor drops.
- Move down to the horizontal axis to find out in which years this occurred.
- So, the amount of money raised stayed the same in 2014 and 2015.

Stem-and-Leaf Plots

A **stem-and-leaf plot** displays data in a table in which the first digits are presented as "stems" and the last digits are presented as "leaves." Stem-and-leaf plots are especially useful when there are large amounts of data and we want to see how the data is distributed. Here is a stem-and-plot that shows scores on a recent math test.

Stem	Leaf
9	2 5 7
8	4 6
7	2 4 6 7 8 9
6	1 3 4 8 9
5	0 2 8 8 9

Several facts become obvious in a quick scan of this stem-and-leaf plot. The highest score was 97 and the lowest was 50. The greatest number of students scored in the 70s and the fewest in the 80s. The vast majority of students scored in the 50s through the 70s. If we dig a little deeper, we see that almost half the students scored in the 50s and 60s. Now, we can see why stem-and-leaf plots are such a compact way of displaying and comparing data.

EXAMPLE

▶ See the stem-and-leaf plot below showing the times it took students to complete a racing course.

CHAPTER 12 Data and Statistics

Stem	Leaf
5	9 14
4	23 38 51
3	9 13 17 35 35 41 52 58
2	11 23 33 47
1	55

2 | 11 = 2 minutes and 11 seconds

What was the fastest racing time? What was the slowest racing time?

▶ To find the fastest racing time, look for the stem with the lowest number and then find the leaf with lowest number. To find the slowest racing time, find the stem with the greatest number and then find the leaf with the greatest number.

▶ The fastest racing time was 1:55, or 1 minute and 55 seconds, and the slowest racing time was 5:14, or 5 minutes and 14 seconds.

Remember that each data point is recorded separately in a stem-and-leaf plot, so sometimes the same number shows up more than once in an interval.

EXAMPLE

▶ Look again at the stem-and-leaf plot from the previous example. Which racing time did two runners in the race both achieve?

▶ Scan the leaf data for each stem and look for the number that appears two times.

▶ Two runners ran the race at 3:35, or 3 minutes and 35 seconds.

Identifying Misleading Statistics

If presented incorrectly, statistics can be misleading. Consider, for example, the bar graph showing the number of tickets sold at three movies during one weekend at a local cineplex:

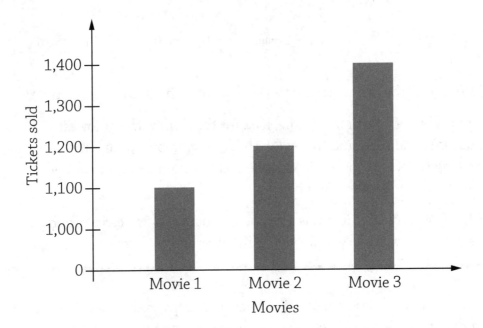

The lengths of the three bars makes it seem as if Movie 3 had more than twice the tickets sold than did Movie 1. In fact, the sales of tickets for Movie 1 were almost 80% of ticket sales for Movie 3: 1,100 ÷ 1,400 = 78.6. The reason the bar graph is deceptive has to do with the vertical scale showing the number of tickets sold. Notice that the intervals on this scale are different. The first interval jumps from 0 to 1,000 whereas other vertical intervals increase by 100. The following line graph shows the growth of a kitten over the first four months of its life.

EXAMPLE

▶ The graph below intends to track a kitten's growth. During which of the first four months did the kitten grow the most? How does this line graph inaccurately represent the data? Explain.

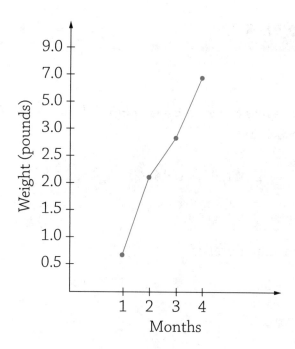

- Examine the intervals on the vertical scale. Notice that they are not uniform. The scale jumps from showing intervals representing $\frac{1}{2}$ pound to intervals representing 2 pounds. This makes it seem that the greatest growth took place between month 1 and month 2.

- If you use the numbers on the vertical scale, you can see that from month 3 to month 4 the kitten's weight increased from 2.75 pounds to 7 pounds.

- So, the kitten grew the most during month 3.

BTW

Broken or unequal intervals are the most common source of misleading data presentations, so always check the labels on the y-axis first.

EXERCISES

EXERCISE 12-1

The numbers of home runs hit by the top ten players during the 2018 baseball season are listed in this data array:

41, 41, 41, 41, 44, 45, 47, 48, 49, 53

1. What is the mean number?

2. What is the median number?

3. What is the mode?

4. What is the range?

EXERCISE 12-2

The pictograph shows the number of baseball games in which four friends played during their summer vacation:

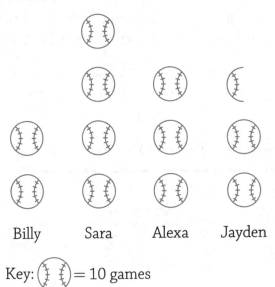

CHAPTER 12   Data and Statistics

1. Which friend played the fewest number of baseball games?

2. How many baseball games did Jayden play?

EXERCISE 12-3

The bar graph shows the number of books read by four sixth-grade classes during the school year.

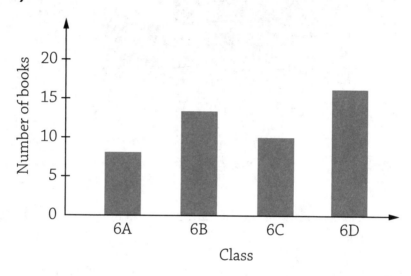

1. How many books did class 6C read during the school year?

2. Which class read about twice as many books as class 6A?

EXERCISE 12-4

The circle graph shows how 6th grade students get to school each day.

Methods of Transportation to School

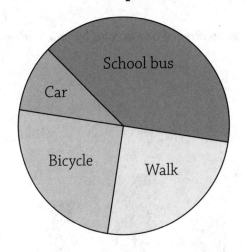

1. What is the least common method for students to get to school each day? How do you know?

2. What is the most common method for students to get to school? Explain.

3. What method do about one in four students use to get to school? How do you know?

EXERCISE 12-5

The line graph shows the high temperatures in Chicago from August 1 to August 7, 2019.

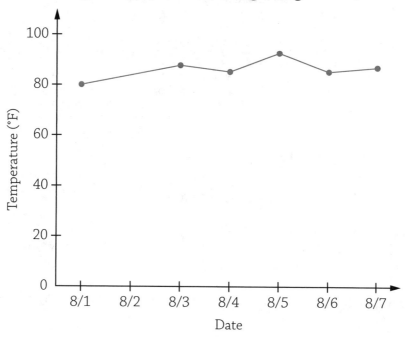

1. On which day was the temperature the highest?

2. About how many degrees warmer was the high temperature on August 5 than on August 1?

EXERCISE 12-6

The stem-and-leaf plot shows the scores of a school's basketball team during the past year.

Stem	Leaf
5	0 3 6
4	4 6 7 9
3	2 5 5 5 9
2	1 3 4 6 7
1	2 4 8

Key: 2 | 4 = 24 points

1. What was the team's most frequent score?

2. What was the team's lowest score?

3. What was the team's highest score?

4. How many points more was the team's highest score than its lowest score?

CHAPTER 12 Data and Statistics 283

EXERCISE 12-7

Identify each type of graph and explain why each graph is misleading.

1. This graph shows the height of the three tallest buildings in a city.

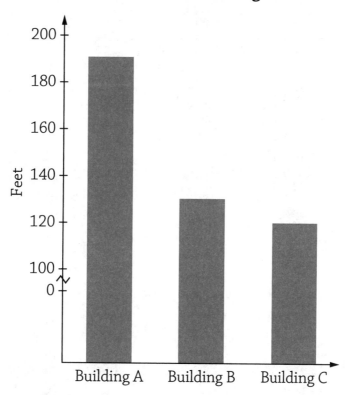

2. This graph shows the scores of the top five skaters at a recent ice skating competition.

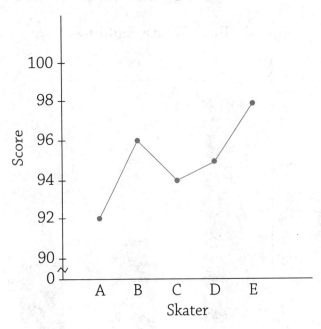

Answer Key

The Number System

EXERCISE 1-1

1. The 4 is in the tens place and its value is four tens or 40.

2. The 6 is in the hundredths place and its value is $\frac{6}{100}$ or 0.06.

EXERCISE 1-2

1.

2.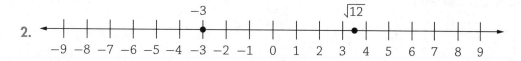

EXERCISE 1-3

1. $\frac{-4}{5}$ is less than $\frac{-5}{7}$ since it is to left of $\frac{-5}{7}$.

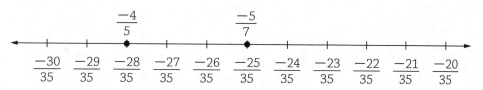

2. From least to greatest: $\frac{-2}{3}$, −0.35, $\frac{1}{5}$, 0.45.

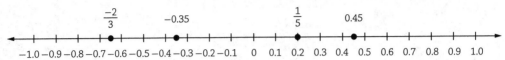

EXERCISE 1-4

1. 8.257 rounded to the nearest hundredth is 8.26.
2. 13.382 rounded to the nearest tenth is 13.4.

EXERCISE 1-5

1.

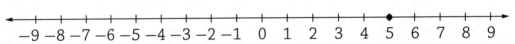

2. 18 is the absolute value of |−18| and |18|.

EXERCISE 1-6

1. 137
2. 61

EXERCISE 1-7

1.
```
    1
   173
    84
 + 211
 ─────
   468
```

2. ${}^{2\,1}$
    $\cancel{3}26$
    -142
    $\overline{184}$

EXERCISE 1-8

1. ${}^{4}$
    16
    $\times7$
    $\overline{112}$

2. 17
    $9\overline{)153}$
    $\underline{-9}$
    63
    $\underline{-63}$
    0

EXERCISE 1-9

1. $6 + 7 = 13$
    $7 + 6 = 13$
    $13 - 6 = 7$
    $13 - 7 = 6$

2. $7 \times 12 = 84$
    $12 \times 7 = 84$
    $84 \div 7 = 12$
    $84 \div 12 = 7$

EXERCISE 1-10

1. commutative property of addition
2. associative property of multiplication
3. distributive property of multiplication over addition
4. inverse property

Decimals

EXERCISE 2-1

1.

Thousands	Hundreds	Tens	Ones	Tenths	Hundredths	Thousands
	3	1	2 .	6	4	9

2.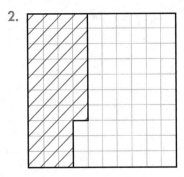

EXERCISE 2-2

1. seventy one and four hundred twenty six thousandths
2. 35.019

3. $6 \times 100 + 4 \times 10 + 9 \times 1 + 3 \times \dfrac{1}{10} + 2 \times \dfrac{1}{100} + 8 \times \dfrac{1}{1000}$

4. $7 \times \dfrac{1}{10} + 4 \times \dfrac{1}{100} + 3 \times \dfrac{1}{1000}$

EXERCISE 2-3

1. $0.3194 > 0.3164$

2.
    1.52 1.56   1.69 1.72
  1.50 1.55 1.60 1.65 1.70 1.75 1.80

3. $6.101 > 6.011 > 6.010 > 6.001$

4. $3.22 < 3.23 < 3.25 < 3.33$

EXERCISE 2-4

1. 5.1
2. 2.75
3. 0.008
4. 0.0004

EXERCISE 2-5

1. The sum of $0.53 + 0.26 = 0.79$. See how the following grid is filled in:

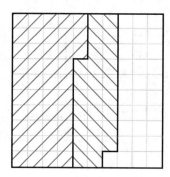

2. The sum of 5.712 + 12.18 = 17.892.

   $$\begin{array}{r} 5.712 \\ +\ 12.180 \\ \hline 17.892 \end{array}$$

3. Mikela jogged a total of 18.879 km.

   $$\begin{array}{r} 6.437 \\ 5.362 \\ +\ 7.180 \\ \hline 18.979 \end{array}$$

4. Rashid collected 83.1 kg of trash.

   $$\begin{array}{r} 35.73 \rightarrow 35.7 \\ +\ 47.39 \rightarrow 47.4 \\ \hline 83.1 \end{array}$$

EXERCISE 2-6

1. The difference between 0.87 and 0.53 is 0.34. See the shaded areas in this grid:

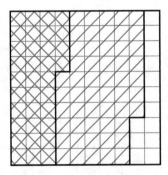

2. The difference between 17 and 10.825 is 6.175.

   $$\begin{array}{r} 17.000 \\ -\ 10.825 \\ \hline 6.175 \end{array}$$

3. Ben will have $4.76 after this purchase.

    $43.75
    − $38.99
    $4.76

4. Juanita will have $2,100.37 on the 15th of the month.

    $2,385.77   Checks: $322.85
    − $1,197.85   + $875.00
    $1187.92 $1,197.85
    + $912.45
    $2,100.37   Deposit: $912.45

EXERCISE 2-7

1. The product of 7 and 0.45 is 3.15.

    0.45
    ×   7
    3.15

2. The product of 4.2 and 0.65 is 2.73.

    0.65
    × 4.20
    000
    130
    260
    2.7300 = 2.73

3. Jocelyn paid $36.25 for the cookies.

    14.50
    ×   2.5
    7250
    2900
    36.250 = 36.25

4. The ant can walk 75.735 inches in 2.25 seconds.

$$
\begin{array}{r}
33.66 \\
\times\ 2.25 \\
\hline
16830 \\
6732 \\
6732 \\
\hline
75.7350 = 75.735
\end{array}
$$

EXERCISE 2-8

1. $487.32 \div 12 = 40.61$

$$
\begin{array}{r}
40.61 \\
12\overline{)487.32} \\
-48 \\
\hline
073 \\
-72 \\
\hline
12 \\
-12 \\
\hline
0
\end{array}
$$

2. $0.702 \div 0.78 = 0.9$

$$0.78\overline{)0.702}$$

↓

$$
\begin{array}{r}
0.9 \\
78\overline{)70.2} \\
-702 \\
\hline
0
\end{array}
$$

3. Each hiker owes $12.85:

```
        12.85
    8)102.80
     −8
      22
     −16
       68
      −64
        40
       −40
         0
```

4. Marcia will have 26 pieces of ribbon:

```
          26
     25)650
       −50
        150
       −150
          0
```

3
Fractions

EXERCISE 3-1

1. $\dfrac{3}{7}$

2. $\dfrac{5}{9}$

EXERCISE 3-2

1. 36

   Multiples of 9: 9, 18, 27, **36**

   Multiples of 12: 12, 24, **36**

2. 42

   Multiples of 14: 14, 28, **42**

   Multiples of 21: 21, **42**

EXERCISE 3-3

1. $2 \times 2 \times 3 \times 5$, or $2^2 \times 3 \times 5$
2. $2 \times 2 \times 2 \times 2 \times 11$, or $2^4 \times 11$

EXERCISE 3-4

1. 3

   Factors of 12: 1, 2, **3**, 4, 6, 12

   Factors of 15: 1, **3**, 5, 15

2. 30

   Factors of 30: 1, 2, 3, 5, 6, 10, 15, **30**

   Factors of 90: 1, 2, 3, 5, 6, 9, 10, 15, 18, **30**, 45, 90

EXERCISE 3-5

1. $\dfrac{1}{24}$

   $\dfrac{1}{6} = \dfrac{1 \times 4}{6 \times 4} = \dfrac{4}{24}$

   $\dfrac{1}{8} = \dfrac{1 \times 3}{8 \times 3} = \dfrac{3}{24}$

Answer Key

$$\frac{4}{24} - \frac{3}{24} = \frac{1}{24}$$

2. $\frac{13}{28}$

$$\frac{3}{4} = \frac{3 \times 7}{4 \times 7} = \frac{21}{28}$$

$$\frac{2}{7} = \frac{2 \times 4}{7 \times 4} = \frac{8}{28}$$

$$\frac{21}{28} - \frac{8}{28} = \frac{13}{28}$$

3. $\frac{9}{10}$ inch

$$\frac{2}{5} = \frac{2 \times 2}{5 \times 2} = \frac{4}{10}$$

$$\frac{1}{2} = \frac{1 \times 5}{2 \times 5} = \frac{5}{10}$$

$$\frac{4}{10} + \frac{5}{10} = \frac{9}{10}$$

4. $\frac{4}{15}$ inch

$$\frac{3}{5} = \frac{3 \times 3}{5 \times 3} = \frac{9}{15}$$

$$\frac{1}{3} = \frac{1 \times 5}{3 \times 5} = \frac{5}{15}$$

$$\frac{9}{15} - \frac{5}{15} = \frac{4}{15}$$

EXERCISE 3-6

1. $8\dfrac{39}{40}$

   $5\dfrac{3}{8} = \dfrac{43}{8}$

   $3\dfrac{3}{5} = \dfrac{18}{5}$

   $\dfrac{43 \times 5}{8 \times 5} = \dfrac{215}{40}$

   $\dfrac{18 \times 8}{5 \times 8} = \dfrac{144}{40}$

   $\dfrac{215}{40} + \dfrac{144}{40} = \dfrac{359}{40} = 8\dfrac{39}{40}$

2. $7\dfrac{5}{6}$ grams

   $5\dfrac{1}{2} = \dfrac{11}{2}$

   $2\dfrac{1}{3} = \dfrac{7}{3}$

   $\dfrac{11 \times 3}{2 \times 3} = \dfrac{33}{6}$

   $\dfrac{7 \times 2}{3 \times 2} = \dfrac{14}{6}$

Answer Key   297

$$\frac{33}{6} + \frac{14}{6} = \frac{47}{6} = 7\frac{5}{6}$$

3. $2\frac{1}{10}$

$$4\frac{1}{2} = \frac{9}{2}$$

$$2\frac{2}{5} = \frac{12}{5}$$

$$\frac{9 \times 5}{2 \times 5} = \frac{45}{10}$$

$$\frac{12 \times 2}{5 \times 2} = \frac{24}{10}$$

$$\frac{45}{10} - \frac{24}{10} = \frac{21}{10} = 2\frac{1}{10}$$

4. $2\frac{2}{15}$

$$4\frac{4}{5} = \frac{24}{5}$$

$$2\frac{2}{3} = \frac{8}{3}$$

$$\frac{24 \times 3}{5 \times 3} = \frac{72}{15}$$

$$\frac{8 \times 5}{3 \times 5} = \frac{40}{15}$$

$$\frac{72}{15} - \frac{40}{15} = \frac{32}{15} = 2\frac{2}{15}$$

EXERCISE 3-7

1. $\dfrac{3}{10}$

    $\dfrac{3}{8} \times \dfrac{4}{5} = \dfrac{12}{40} = \dfrac{3}{10}$

2. $\dfrac{8}{27}$

    $\dfrac{2}{9} \div \dfrac{3}{4} = \dfrac{2}{9} \times \dfrac{4}{3} = \dfrac{8}{27}$

3. $\dfrac{1}{2}$ mile

    $\dfrac{3}{4} \times \dfrac{2}{3} = \dfrac{6}{12} = \dfrac{1}{2}$

4. 10 bags

    $2 \div \dfrac{1}{5} = \dfrac{2}{1} \times \dfrac{5}{1} = \dfrac{10}{1} = 10$

EXERCISE 3-8

1. 12

    $5\dfrac{1}{3} = \dfrac{16}{3}$

    $2\dfrac{1}{4} = \dfrac{9}{4}$

    $\dfrac{16}{3} \times \dfrac{9}{4} = \dfrac{144}{12} = 12$

Answer Key

2. $3\dfrac{1}{11}$

$$8\dfrac{1}{2} = \dfrac{17}{2}$$

$$2\dfrac{3}{4} = \dfrac{11}{4}$$

$$\dfrac{17}{2} \div \dfrac{11}{4} = \dfrac{17}{2} \times \dfrac{4}{11} = \dfrac{68}{22} = \dfrac{34}{11} = 3\dfrac{1}{11}$$

3. $6\dfrac{2}{3}$ cups

$$2\dfrac{2}{3} = \dfrac{8}{3}$$

$$2\dfrac{1}{2} = \dfrac{5}{2}$$

$$\dfrac{8}{3} \times \dfrac{5}{2} = \dfrac{40}{6} = 6\dfrac{4}{6} = 6\dfrac{2}{3}$$

4. about 13 days

$$10\dfrac{1}{2} \div \dfrac{4}{5}$$

$$10\dfrac{1}{2} = \dfrac{21}{2}$$

$$\dfrac{21}{2} \div \dfrac{4}{5} = \dfrac{21}{2} \times \dfrac{5}{4} = \dfrac{105}{8} = 13\dfrac{1}{8}$$

5. 5 boxes

$100 \div 18\frac{3}{4}$

$18\frac{3}{4} = \frac{75}{4}$

$100 \div \frac{75}{4} = \frac{100}{1} \times \frac{4}{75} = \frac{400}{75} = 5\frac{25}{75} = 5\frac{1}{3}$

Integers

EXERCISE 4-1

1. $5 > -2$

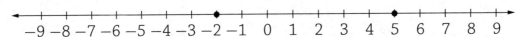

2. 6 is the opposite of -6

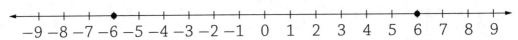

EXERCISE 4-2

1. $-2 + (-7) = -9$

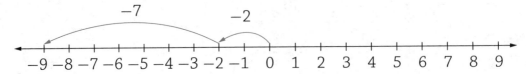

Answer Key   **301**

2. $-4 + 6 = 2$

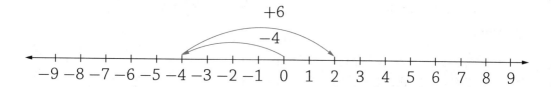

3. The sum of 7 and $-3 = 4$

   $7 + (-3) = |7| - |3| = |4| = 4$

4. 16 miles

   $5 + 4 + 3 + 4 = 16$

EXERCISE 4-3

1. 16

   $5 - (-11) = 5 + 11 = 16$

2. -14

   $-8 - (+6) = -8 - 6 = -14$

3. 17

   $28 - (+11) = 28 - 11 = 17$

4. 655 feet

   $1{,}650 - 550 - 445 = 655$

EXERCISE 4-4

1. 544

   $-32 \times -17 = 544$

2. -60

   $-15 \times 4 = -60$

3. −468

$-4 \times -9 \times -13 = [-4 \times -9] \times -13 = 36 \times -13 = -468$

4. 21°F

$-3°F \text{ per hour} \times 5 \text{ hours} = -15°F$
→ $36°F - 15°F = 21°F$

EXERCISE 4-5

1. 5

$-35 \div -7 = 5$

2. −7

$63 \div -9 = -7$

3. 13

$-156 \div -12 = 13$

4. −$2.50

$-\$12.50 \div 5 = -\$2.50$

5

Ratio and Proportion

EXERCISE 5-1

1. $\frac{5}{8}$ is the simplest form of the ratio of computers to students.

$$\frac{25}{40} = \frac{25 \div 5}{40 \div 5} = \frac{5}{8}$$

Answer Key 303

2. The ratio of show minutes to advertising minutes is $\frac{13}{7}$.

$$\frac{39 \text{ show minutes}}{21 \text{ advertising minutes}} = \frac{13}{7}$$

3. The ratio of girls to boys in the afterschool program in simplest form is $\frac{11}{12}$.

$$\frac{22 \text{ girls}}{24 \text{ boys}} = \frac{11}{12}$$

4. The ratio of red roses to yellow roses in simplest form is $\frac{7}{4}$.

$$\frac{28 \text{ red}}{16 \text{ yellow}} = \frac{7}{4}$$

EXERCISE 5-2

1. Yes, the mid-season ratio of wins to games played is the same as the end-of-season ratio of wins to games played. The cross products of the ratios $\frac{8}{14}$ and $\frac{12}{21}$ are equal, so they are equivalent ratios.

$$\frac{8}{14} = \frac{2}{21}$$
$$8 \times 21 = 14 \times 12$$
$$168 = 168$$

2. = equal to
3. > greater than
4. < less than
5. = equal to

Answer Key

EXERCISE 5-3

1. No, the ratio of right-handed students is greater in Mr. Jessup's class than in the ratio of right-handed students in the entire sixth grade.

$$\frac{23}{25} \; ? \; \frac{110}{125}$$

   Multiply the cross products: $23 \times 125 = 2{,}875$ and $25 \times 110 = 2{,}750$. Since the cross products are not equal, the ratios are not equivalent.

2. Caleb's collection has the greater ratio of New York Knicks to Miami Heat cards.

   Caleb's collection: $\dfrac{\text{Knicks}}{\text{Heat}} = \dfrac{23}{17} \approx 1.353$

   Malik's collection: $\dfrac{\text{Knicks}}{\text{Heat}} = \dfrac{24}{18} \approx 1.333$

   $1.353 > 1.333$

EXERCISE 5-4

1. The Judson's SUV gets 25 miles per gallon of gasoline.

   $$\frac{\text{miles}}{\text{gallons}} = \frac{325 \div 13}{13 \div 13} = 25$$

2. Rachel makes more widgets per hour compared to Juan.

   Rachel: $\dfrac{\text{widgets}}{\text{hours}} = \dfrac{210}{3} = 70$ widgets per hour

   Juan: $\dfrac{\text{widgets}}{\text{hours}} = \dfrac{260}{4} = 65$ widgets per hour

EXERCISE 5-5

1. $x = 13$

    $$\frac{4}{12} = \frac{x}{39}$$
    $$12x = 156$$
    $$x = 13$$

2. $k = 2$

    $$\frac{k}{15} = \frac{6}{45}$$
    $$45k = 90$$
    $$k = 2$$

3. About 1,333 cell phones out of 200,000 newly manufactured cell phones would be defective.

    $$\frac{\text{defective phones}}{\text{total phones made}} = \frac{2}{300} = \frac{x}{200{,}000}$$
    $$300x = 400{,}000$$
    $$x \approx 1{,}333$$

4. You sold 60 raffle tickets.

    $$\frac{12}{5} = \frac{x}{25}$$
    $$5x = 300$$
    $$x = 60$$

Percent

EXERCISE 6-1

1. $\frac{3}{5}$, 0.6, 60%

2. $\frac{1}{4}$, 0.25, 25%

EXERCISE 6-2

1. 15%

    Sections C and D combined make up 25% of the circle, and section C is bigger than Section D. Half of 25% is 12.5%, so Section C must be slightly more than 12.5%. Therefore, 15% seems like a reasonable estimate.

2. 30%

    The bottom half of the rectangle represents 50% of the figure. Slightly more than half of the bottom is unshaded. Since half of the bottom represent 25%, the unshaded portion is somewhat greater. Therefore, 30% seems like a reasonable estimate.

EXERCISE 6-3

1. 0.76

    $19 \div 25 = 0.76$

2. 65%

    $13 \div 20 = 0.65 = 65\%$

3. $\dfrac{45}{100}$ or $\dfrac{9}{20}$

4. 0.55

EXERCISE 6-4

1. 147

    $2{,}100 \times 0.07 = 147$

2. $700

    $\$500 \times 1.4 = \$700$

3. 20%

    $x\% \times 375 = 75$

    $$x = \dfrac{75}{375}$$
    $$= 0.20$$
    $$= 20\%$$

4. 82%

    $x\% \times 85 = 70$

    $$x = \dfrac{70}{85}$$
    $$= 0.823$$
    $$\approx 82\%$$

5. 1,400

    $0.55 \times x = 770$

    $$x = \dfrac{770}{0.55}$$
    $$x = 1{,}400$$

6. 84 cell phones

$$1.25x = 105$$
$$\frac{1.25x}{1.25} = \frac{105}{1.25}$$
$$x = 84$$

EXERCISE 6-5

1. 37.5%

    Amount of decrease: $400 - $250 = $150

    Percent of decrease: $\frac{150}{400} = 0.375 = 37.5\%$

2. 12.5%

    Amount of increase: $45 - $40 = $5

    Percent of increase: $\frac{5}{40} = 0.125 = 12.5\%$

EXERCISE 6-6

1. $800

    $$I = \$10{,}000 \times 0.02 \times 4$$
    $$I = \$800$$

2. $4,800

    $$I = \$20{,}000 \times 0.08 \times 3$$
    $$I = \$4{,}800$$

Answer Key 309

7
Equations and Inequalities

EXERCISE 7-1

1. $n + 4$
2. $8n$
3. $n - 2$
4. $n \div 2$ or $\dfrac{n}{2}$

EXERCISE 7-2

1. $x = 8$

    Solution:
    $$x + 7 = 15$$
    $$x + 7 - 7 = 15 - 7$$
    $$x = 8$$

    Check:
    $$x + 7 = 15$$
    $$8 + 7 = 15$$
    $$15 = 15$$

2. $y = 9$

    Solution:
    $$6y = 54$$
    $$\dfrac{6y}{6} = \dfrac{54}{6}$$
    $$y = 9$$

Check:
$$6y = 54$$
$$6 \times 9 = 54$$
$$54 = 54$$

3. 490 ft

    cable length = building height + antennae height + underground depth

    Solution/Check:
    $$C = B + A + U$$
    $$C = 400 + 75 + 15$$
    $$C = 490$$

4. $x = 15$
    $$\frac{3x}{5} - 6 = 3$$
    $$\frac{3x}{5} - 6 + 6 = 3 + 6$$
    $$\frac{3x}{5} = 9$$
    $$3x = 45$$
    $$\frac{3x}{3} = \frac{45}{3}$$
    $$x = 15$$

    Check:
    $$\frac{3}{5}x - 6 = 3$$
    $$\frac{3}{5}(15) - 6 = 3$$
    $$\frac{45}{5} - 6 = 3$$
    $$9 - 6 = 3$$
    $$3 = 3$$

EXERCISE 7-3

1.

x	y	(x, y)
−2	3	(−2, 3)
0	5	(0, 5)
2	7	(2, 7)

2.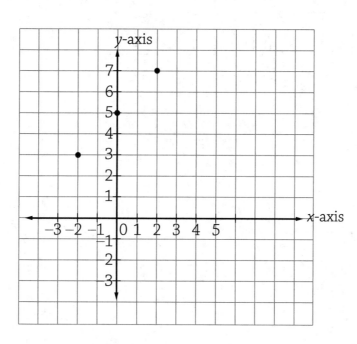

3.

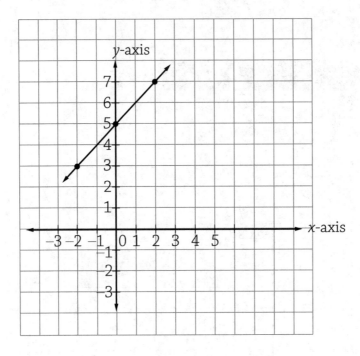

4. 6

EXERCISE 7-4

1. a monomial
2. a polynomial or trinomial
3. $4x^4 + 2x^2 + 5x - 13$
4. fourth degree

EXERCISE 7-5

1. $7x^2 + 6x + 3$

$$\begin{array}{r} 5x^2 + 3x - 2 \\ +2x^2 + 3x + 5 \\ \hline 7x^2 + 6x + 3 \end{array}$$

2. $6y^2 - 4y + 4$

$$\begin{aligned} 4y^2 - y + 8 \\ + 2y^2 + 3y - 4 \\ \hline 6y^2 - 4y + 4 \end{aligned}$$

3. $10x^5 + 8x^2$

$(2x^2) \times (5x^3 + 4)$
$= (2x^2 \times 5x^3) + (2x^2 \times 4)$
$= 10x^5 + 8x^2$

4. $4k^6 + 44k^3$

$(4k^3) \times (k^3 + 11)$
$= (4k^3 \times k^3) + (4k^3 \times 11)$
$= 4k^6 + 44k^3$

EXERCISE 7-6

1. $x \times 4$

$$x + 8 \geq 12$$
$$x + 8 - 8 \geq 12 - 8$$
$$x \geq 4$$

2. $x \leq 15$

$$\frac{2}{5}x \leq 6$$
$$\frac{5}{2} \times \frac{2}{5}x \leq \frac{5}{2} \times 6$$
$$x \leq \frac{30}{2}$$
$$x \leq 15$$

3. $y \leq 7$

$$8y \leq 56$$
$$\frac{8y}{8} \leq \frac{56}{8}$$
$$y \leq 7$$

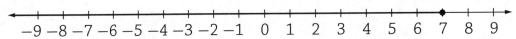

4. $n > 4$

$$5n - 10 > 10$$
$$5n - 10 + 10 > 10 + 10$$
$$5n > 20$$
$$\frac{5n}{5} > \frac{20}{5}$$
$$n > 4$$

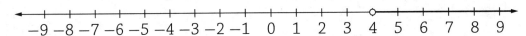

Measurement and Geometry

EXERCISE 8-1

1. 15 ft

    $3 \times 5 = 15$

2. 36 in

    $12 \times 3 = 36$

3. 3,520 yd

   $1{,}760 \times 2 = 3{,}520$

4. 7,920 ft

   $5{,}280 \times 1.5 = 7{,}920$

EXERCISE 8-2

1. 4.5 lb

   $72 \div 16 = 4.5$

2. 96 oz

   $16 \times 6 = 96$

3. 10,000 lb

   $2{,}000 \times 5 = 10{,}000$

4. 2.25 T

   $4{,}500 \div 2{,}000 = 2.25$

EXERCISE 8-3

1. 6 c

   $2 \times 3 = 6$

2. 16 fl oz

   $8 \times 2 = 16$

3. 12 qt

   $4 \times 3 = 12$

4. 4 pt

   $2 \times 2 = 4$

EXERCISE 8-4

1. 210 min
   $60 \times 3.5 = 210$
2. 168 hr
   $24 \times 7 = 168$
3. 900 s
   $60 \times 15 = 300$
4. 250 yr
   $100 \times 2.5 = 250$
5. 520 weeks
   $52 \times 10 = 520$
6. 5 yr
   $60 \div 12 = 5$

EXERCISE 8-5

1. 300 cm
   $100 \times 3 = 300$
2. 1,500 mm
   $1,000 \times 1.5 = 1,500$
3. 7.5 km
   $7,500 \div 1,000 = 7.5$
4. 15 dm
   $10 \times 1.5 = 15$
5. 5,000 m
   $1,000 \times 5 = 5,000$

Answer Key   317

EXERCISE 8-6

1. milligrams
2. kilograms or grams
3. grams
4. 4 melons: 750 g × 4 = 3,000 g = 3 kg
5. 2,500 mg: 1,000 × 2.5 = 2,500

EXERCISE 8-7

1. milliliter
2. liter
3. 6 L
   6,000 ÷ 1,000 = 6
4. 2.1 L
   350 × 6 = 2,100; 2,100 ÷ 1,000 = 2.1

EXERCISE 8-8

1. 18 cm
   4 + 5 + 4 + 5 = 18
2. 12 in
   5 + 4 + 3 = 12
3. 30 ft
   5 + 4 + 5 + 3 + 8 + 5 = 30
4. 14 m
   2 + 2 + 4 + 2 + 2 + 2 = 14

EXERCISE 8-9

1. 21.98 cm

$$\begin{aligned} C &= \pi d \\ &= 3.14 \times 7 \\ &= 21.98 \end{aligned}$$

2. 18.86 in

$$\begin{aligned} C &= 2\pi r \\ &= (2 \times 3) \times \left(\frac{22}{7}\right) \\ &= 6\left(\frac{22}{7}\right) \\ &= \frac{132}{7} \\ &= 18.86 \end{aligned}$$

Plane Geometry

EXERCISE 9-1

1. This is a line segment and is written as $\overline{RS}$.
2. This is a ray and is written as $\overrightarrow{PQ}$.
3. This is a line and is written as $\overleftrightarrow{LM}$.
4. This is a ray and is written as $\overrightarrow{TU}$.

EXERCISE 9-2

1. acute angle
2. obtuse angle
3. straight angle
4. right angle

EXERCISE 9-3

1. obtuse triangle
2. acute triangle
3. right triangle

EXERCISE 9-4

1. equilateral triangle
2. scalene triangle
3. isosceles triangle

EXERCISE 9-5

1. $m\angle x = 180 - (39 + 42)$
   $= 180 - 81$
   $= 99°$
2. $m\angle x = 180 - (32 + 85)$
   $= 180 - 117$
   $= 63°$

EXERCISE 9-6

1. trapezoid
2. rectangle
3. parallelogram
4. rhombus

EXERCISE 9-7

1. $m\angle x = 360 - (135 + 105 + 60)$
   $= 360 - 300$
   $= 60°$
2. $m\angle x = 360 - (103 + 82 + 90)$
   $= 360 - 275$
   $= 85°$

EXERCISE 9-8

1. octagon
2. hexagon
3. pentagon
4. heptagon

EXERCISE 9-9

1. $S = (n - 2)(180°)$
   $= (6 - 2)180$
   $= (4)180$
   $= 720°$
2. $S = (n - 2)(180°)$
   $= (7 - 2)180$
   $= (5)180$
   $= 900°$

EXERCISE 9-10

1. $900° \div 7 \approx 128.6°$
2. $720° \div 6 = 120.0°$

EXERCISE 9-11

1. neither
2. congruent
3. similar

EXERCISE 9-12

1. $S = (n - 2)(180°)$

   $= (4 - 2)(180°)$

   $= 2(180°) = 360°$

   The sum of the interior angles is 360°.

   $m\angle x = 360 - (45 + 90 + 105)$

   $= 360 - 240$

   $= 120°$

   $\angle x$ measures 120°.

2. $S = (n - 2)(180°)$

   $= (5 - 2)(180°)$

   $= 3(180°) = 540°$

   The sum of the interior angles is 540°.

   $m\angle x = 540 - (90 + 120 + 90 + 115)$

$$= 540 - 415$$
$$= 125°$$

So, ∠x measures 125°.

EXERCISE 9-13

1. The figure has two lines of symmetry.

2. The figure has four lines of symmetry.

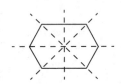

EXERCISE 9-14

1.

2.

Answer Key 323

10
Geometry: Area and Volume

EXERCISE 10-1

1. 25 cm². Use the formula for the area of a square: $A = s^2$.
   $A\text{(tile)} = 5 \text{ cm} \times 5 \text{ cm} = 25 \text{ cm}^2$

2. 875 ft². Use the formula for the area of a rectangle: $A = l \times w$.
   $A\text{(floor)} = 35 \text{ ft} \times 25 \text{ ft} = 875 \text{ ft}^2$

EXERCISE 10-2

1. 21 in². Use the formula for the area of a parallelogram: $A = b \times h$.
   $A = 7 \text{ in} \times 3 \text{ in} = 21 \text{ in}^2$

2. 72 cm². Use the formula for the area of a parallelogram: $A = b \times h$.
   $A = 12 \text{ cm} \times 6 \text{ cm} = 72 \text{ cm}^2$

EXERCISE 10-3

1. 60 in². Use the formula for the area of a triangle: $A = \frac{1}{2} \times b \times h$.
   $A = \frac{1}{2} \times 15 \text{ in} \times 8 \text{ in} = 60 \text{ in}^2$

2. 15 km². Use the formula for the area of a triangle: $A = \frac{1}{2} \times b \times h$.
   $A = \frac{1}{2} \times 6 \text{ km} \times 5 \text{ km} = 15 \text{ km}^2$

3. 140.4 in². Use the formula for the area of a triangle: $A = \frac{1}{2} \times b \times h$.
   $A\text{(yield sign)} = \frac{1}{2} \times 18 \text{ in} \times 15.6 \text{ in} = 140.4 \text{ in}^2$

4. 21 cm². Use the formula for the area of a triangle: $A = \frac{1}{2} \times b \times h$.

   $A(\text{patch}) = \frac{1}{2} \times 6 \text{ cm} \times 7 \text{ cm} = 21 \text{ cm}^2$

EXERCISE 10-4

1. 201 cm². Use the formula for the area of a circle: $A = \pi \times r^2$.

   $A = 3.14 \times 8 \text{ cm} \times 8 \text{ cm} = 200.96 \text{ cm}^2 \approx 201 \text{ cm}^2$

2. 615.4 in². Use the formula for the area of a circle: $A = \pi \times r^2$. The diameter of the fountain is 28 in, so its radius is 14 in.

   $A = 3.14 \times 14 \text{ in} \times 14 \text{ in} = 615.4 \text{ in}^2$

3. 1,384.7 cm². Use the formula for the area of a circle: $A = \pi \times r^2$. The diameter of the dartboard is 42 cm, so its radius is 21 cm.

   $A = 3.14 \times 21 \text{ cm} \times 21 \text{ cm} = 1{,}384.74 \text{ cm}^2 \times 1{,}384.7 \text{ cm}^2$

4. 314 in². Use the formula for the area of a circle: $A = \pi \times r^2$. Note that the radius of each photograph is 5 in and there are four circles.

   $A = 3.14 \times 5 \text{ in} \times 5 \text{ in} \times 4 = 314 \text{ in}^2$

EXERCISE 10-5

1. 66 cm². Use the formula for the surface area of a rectangular prism:

   $SA = 2lw + 2lh + 2wh$

   $SA(\text{top/bottom of prism}, 2lw) = 2 \times 3 \text{ cm} \times 3 \text{ cm} = 18 \text{ cm}^2$

   $SA(\text{sides of prism}, 2lh) = 2 \times 3 \text{ cm} \times 4 \text{ cm} = 24 \text{ cm}^2$

   $SA(\text{opposite sides of prism}, 2wh) = 2 \times 3 \text{ cm} \times 4 \text{ cm} = 24 \text{ cm}^2$

   $SA(\text{prism}) = 18 \text{ cm}^2 + 24 \text{ cm}^2 + 24 \text{ cm}^2 = 66 \text{ cm}^2$

2. 104 cm². Use the formula for the surface area of a rectangular prism:

$SA = 2lw + 2lh + 2wh$

SA(top/bottom of box, $2lw$) = 2 × 6 cm × 5 cm = 60 cm²

SA(long side of box, $2lh$) = 2 × 6 cm × 2 cm = 24 cm²

SA(short side of box, $2wh$) = 2 × 5 cm × 2 cm = 20 cm²

SA(box) = 60 cm² + 24 cm² + 20 cm² = 104 cm²

EXERCISE 10-6

1. 2,198 in². Use the formula for the surface area of a cylinder: $SA = 2\pi rh + 2\pi r^2$. Notice that the diameter of the cylinder is 20 in, so its radius is 10 in.

SA(cylinder face, $2\pi rh$) = 2 × 3.14 × 10 in × 25 in = 1,570 in²

SA(cylinder bases, $2\pi r^2$) = 2 × 3.14 × 10 in × 10 in = 628 in²

SA(cylinder) = 1,570 in² + 628 in² = 2,198 in²

2. 7,787.2 mm². Use the formula for the surface area of a cylinder: $SA = 2\pi rh + 2\pi r^2$. Notice that the diameter of the battery is 32 mm, so its radius is 16 mm.

SA(battery face, $2\pi rh$) = 2 × 3.14 × 16 mm × 61.5 mm = 6,179.52 mm²

SA(battery bases, $2\pi r^2$) = 2 × 3.14 × 16 mm × 16 mm = 1,607.68 mm²

SA(battery) = 6,179.52 mm² + 1,607.68 mm² = 7,787.2 mm²

EXERCISE 10-7

1. 251.2 in². Use the formula for the surface area of a cone: $SA = \pi rs + \pi r^2$.
   Notice that the diameter of the wax candle is 10 in, so its radius is 5 in.
   SA(candle face, πrs) = 3.14 × 5 in × 11 in = 172.7 in²
   SA(candle base, πr^2) = 3.14 × 5 in × 5 in = 78.5 in²
   SA(candle) = 172.7 in² + 78.5 in² = 251.2 in²

2. 113 in². Use the formula for the surface area of a cone: $SA = \pi rs + \pi r^2$.
   SA(cone face, πrs) = 3.14 × 4 in × 5 in = 62.8 in²
   SA(cone base, πr^2) = 3.14 × 4 in × 4 in = 50.24 in²
   SA(cone) = 62.8 in² + 50.24 in² = 113.04 in² ≈ 113 in²

EXERCISE 10-8

1. 201 cm². Use the formula for the surface area of a sphere: $SA = 4\pi r^2$.
   Notice that the diameter of the sphere is 8 cm, so its radius is 4 cm.
   SA = 4 × 3.14 × 4 cm × 4 cm = 200.96 cm² ≈ 201 cm²

2. 615.4 in². Use the formula for the surface area of a sphere: $SA = 4\pi r^2$.
   SA = 4 × 3.14 × 7 in × 7 in = 615.44 in² ≈ 615.4 in²

EXERCISE 10-9

1. 82.9 m³. Use the formula for the volume of a rectangular prism: $V = lwh$.
   V = 6.5 m × 2.2 m × 5.8 m = 82.94 m³ ≈ 82.9 m³

2. 54 ft³. Use the formula for the volume of a rectangular prism: $V = lwh$.
   V(flower box) = 9 ft × 3 ft × 2 ft = 54 ft³

EXERCISE 10-10

1. 768 cm³. Use the formula for the volume of a rectangular prism: $V = lwh$. To solve the problem, break the figure into two rectangular prisms. Use the adjusted measures as shown below.

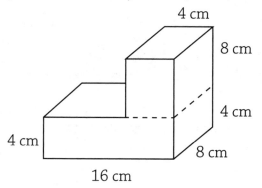

V(bottom prism) = 16 cm × 8 cm × 4 cm = 512 cm³

V(top prism) = 4 cm × 8 cm × 8 cm = 256 cm³

V(irregular rectangular figure) = 512 cm³ + 256 cm³ = 768 cm³

2. 288 in³. Use the formula for the volume of a rectangular prism: $V = lwh$. To solve the problem, break the figure into two rectangular prisms. Use the adjusted measures as shown below.

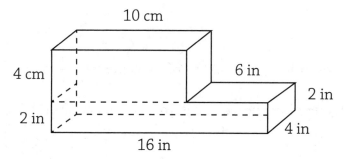

V(bottom prism) = 16 in × 4 in × 2 in = 128 in³

V(top prism) = 10 in × 4 in × 4 in = 160 in³

V(irregular rectangular figure) = 128 in.³ + 160 in.³ = 288 in³

EXERCISE 10-11

1. 502.4 m³. Use the formula for the volume of a cylinder: $V = \pi r^2 h$. Notice that the diameter of the cylinder is 8 m, so its radius is 4 m.

   $V = 3.14 \times 4 \text{ m} \times 4 \text{ m} \times 10 \text{ m} = 502.4 \text{ m}^3$

2. 50.2 m³. Use the formula for the volume of a cylinder: $V = \pi r^2 h$.

   $V = 3.14 \times 2 \text{ m} \times 2 \text{ m} \times 4 \text{ m} = 50.24 \text{ m}^3 \approx 50.2 \text{ m}^3$

EXERCISE 10-12

1. 100.5 in³. Use the formula for the volume of a cone: $V = \frac{1}{3}(\pi r^2 h)$.

   $V = \frac{1}{3}(3.14 \times 4 \text{ in} \times 4 \text{ in} \times 6 \text{ in})$

   $V = \frac{1}{3}(301.44 \text{ in}^3) = 100.48 \text{ in}^3 \approx 100.5 \text{ in}^3$

2. 733 in³. Use the formula for the volume of a cone: $V = \frac{1}{3}(\pi r^2 h)$. Notice that the diameter of the cone is 10 in, so its radius is 5 in.

   $V = \frac{1}{3}(3.14 \times 5 \text{ in} \times 5 \text{ in} \times 28 \text{ in})$

   $V = \frac{1}{3}(2{,}198 \text{ in}^3) \approx 733 \text{ in}^3$

EXERCISE 10-13

1. 2,143.6 in³. Use the formula for the volume of a sphere: $V = \frac{4}{3}\pi r^3$.

   Notice that the diameter of the beach ball is 16 in., so its radius is 8 in.

   $V = \frac{4}{3}(3.14 \times 8 \text{ in} \times 8 \text{ in} \times 8 \text{ in})$

   $V = \frac{4}{3}(1{,}607.68 \text{ in}^3) \approx 2{,}143.57 \text{ in}^3 \approx 2{,}143.6 \text{ in}^3$

2. 33.5 mm³. Use the formula for the volume of a sphere: $V = \frac{4}{3}\pi r^3$. Notice that the diameter of the steel ball is 4 mm, so its radius is 2 mm.

$V(\text{steel ball}) = \frac{4}{3}(3.14 \times 2 \text{ mm} \times 2 \text{ mm} \times 2 \text{ mm})$

$V(\text{steel ball}) = \frac{4}{3}(25.12 \text{ mm}^3) \approx 33.49 \text{ mm}^3 \approx 33.5 \text{ mm}^3$

11
Probability

EXERCISE 11-1

1. $P(\text{blue}) = \frac{2}{10} = 20\%$
2. $P(\text{consonant}) = \frac{6}{8} = 0.75$
3. $P(\text{letter I}) = \frac{0}{8} = 0$
4. $P(\text{greater than 8}) = \frac{4}{12} = \frac{1}{3}$

EXERCISE 11-2

1.

Penny	Nickel	Outcome
Heads	Heads	Heads/Heads
	Tails	Heads/Tails
Tails	Heads	Tails/Heads
	Tails	Tails/Tails

There are four possible outcomes.

2.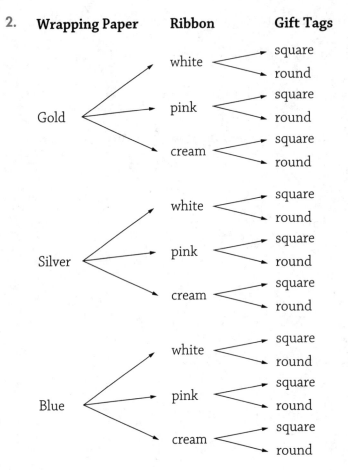

There are 18 possible outcomes.

EXERCISE 11-3

1.

Choice A Tenor 1	Choice B Tenor 2
Abe	Ben
Abe	Charles
~~Ben~~	~~Abe~~
Ben	Charles
~~Charles~~	~~Abe~~
~~Charles~~	~~Ben~~

There are three possible combinations of two tenors that can be selected.

2.

Main Course 1 → Drink 1
Main Course 1 → Drink 2

Main Course 2 → Drink 1
Main Course 2 → Drink 2

Main Course 3 → Drink 1
Main Course 3 → Drink 2

Main Course 4 → Drink 1
Main Course 4 → Drink 2

There are eight possible combinations of main course and a drink for lunch.

EXERCISE 11-4

1. There are six different permutations: BA, BT, AB, AT, TB, and TA.

2.

Morning Shift	Afternoon Shift
Deb	Ed
Deb	Fran
Deb	George
Ed	Deb
Ed	Fran
Ed	George
Fran	Deb
Fran	Ed
Fran	George
George	Deb
George	Ed
George	Fran

There are 12 different permutations.

EXERCISE 11-5

1. The probability that they both go to the sci-fi movie is $\frac{1}{4}$.

The problem assumes that the events are independent, so the formula to use is $P(A \text{ and } B) = P(A) \times P(B)$.

$$\frac{1}{2} \times \frac{1}{2} = \frac{1}{4}$$

2. The probability you will roll two 6s is $\frac{1}{4}$.

$$P = \frac{3}{6} \times \frac{3}{6} = \frac{9}{36} = \frac{1}{4}$$

3. The probability that the two balls are white and black is $\frac{1}{5}$.

$$P = \frac{1}{2} \times \frac{2}{5} = \frac{2}{10} = \frac{1}{5}$$

4. The probability that both spins are orange is $\frac{1}{16}$.

$$P = \frac{1}{4} \times \frac{1}{4} = \frac{1}{16}$$

EXERCISE 11-6

1. $P = \frac{145}{418} \approx 0.35$

2. $P = \frac{5}{40} = 0.125 = 12.5\%$

12

Data and Statistics

EXERCISE 12-1

1. 45: $450 \div 10 = 45$

2. 44.5: Since the data set has an even number of items, the average of the fifth and sixth items is the median.

   $44 + 45 = 89 \div 2 = 44.5$

3. 41: This number occurs most frequently.

4. 12: $53 - 41 = 12$

EXERCISE 12-2

1. Billy, 20 games played
2. 25 games

EXERCISE 12-3

1. 10 books
2. Class 6D

EXERCISE 12-4

1. By car: This section of the pie graph is the smallest.
2. By school bus: This section of the pie graph is the largest.
3. Walk: This section is about $\frac{1}{4}$ of the circle.

EXERCISE 12-5

1. August 5
2. about 11–12°F

EXERCISE 12-6

1. 35 points
2. 12 points
3. 56 points
4. 44 points: $56 - 12 = 44$

EXERCISE 12-7

1. The break in the scale makes Building A seem twice as tall as Building B, but it is only 60 feet taller.

2. Because of the broken scale and small intervals, it looks as if Skater B performed much better than Skater C when compared to how well Skater D did against Skater C. In fact, the difference in scores between these two sets of skaters is only 1 point.

NOTES

NOTES

NOTES

NOTES

NOTES